LEAVING SPACE FOR NATURE

This book provides the first contemporary assessment of area-based conservation and its implications for nature and society.

Now covering 15 per cent of the land surface and a growing area of ocean, the creation of protected areas is one of the fastest conscious changes in land management in history. But this has come at a cost, including a backlash from human rights organisations about the social impacts of protected areas. At the same time, a range of new types of area-based conservation has emerged, based on indigenous people's territories, local community lands and a new designation of "other effective area-based conservation measures". This book provides a concise overview of the status and possible futures of area-based conservation. With many people calling for half the earth's land surface to remain in a natural condition, this book taps into the urgent debate about the feasibility of such an aim and the ways in which such land might be managed. It provides a timely contribution by people who have been at the centre of the debate for the last twenty years. Building on the authors' large personal knowledge, the book draws on global case studies where the authors have first-hand experience, including Yosemite National Park (USA), Blue Mountains National Park (Australia), Bwindi National Park (Uganda), Chingaza National Park (Colombia), Ustyart Plateau (Kazakhstan), Snowdonia National Park (Wales) and many more.

This book is essential reading for students, academics and practitioners interested in conservation and its impact on society.

Nigel Dudley is a consultant ecologist who has worked with international organisations, including WWF International, IUCN and UNESCO. He is co-founder of Equilibrium Research and an industry fellow in the School of Earth and Environmental Sciences at the University of Queensland, Australia. Nigel is the author/editor of numerous titles, including *Arguments for Protected Areas* (Routledge, 2010) and *Authenticity in Nature* (Routledge, 2011).

Sue Stolton co-founded Equilibrium Research with Nigel Dudley thirty years ago. She works mainly on issues relating to protected areas, including management of protected areas and the wider values and benefits that protected areas offer. Sue is vice chair of the WCPA specialist group on privately protected areas and nature stewardship.

Routledge Studies in Conservation and the Environment

This series includes a wide range of inter-disciplinary approaches to conservation and the environment, integrating perspectives from both social and natural sciences. Topics include, but are not limited to, development, environmental policy and politics, ecosystem change, natural resources (including land, water, oceans and forests), security, wildlife, protected areas, tourism, human-wildlife conflict, agriculture, economics, law and climate change.

Natural Resources, Tourism and Community Livelihoods in Southern Africa
Challenges of Sustainable Development
Edited by Moren T. Stone, Monkgogi Lenao and Naomi Moswete

Leaving Space for Nature
The Critical Role of Area-Based Conservation
Nigel Dudley and Sue Stolton

For more information about this series, please visit: www.routledge.com/Routledge-Studies-in-Conservation-and-the-Environment/book-series/RSICE

LEAVING SPACE FOR NATURE

The Critical Role of Area-Based Conservation

Nigel Dudley and Sue Stolton

Routledge
Taylor & Francis Group

LONDON AND NEW YORK

First published 2020
by Routledge
2 Park Square, Milton Park, Abingdon, Oxon OX14 4RN

and by Routledge
52 Vanderbilt Avenue, New York, NY 10017

Routledge is an imprint of the Taylor & Francis Group, an informa business

British Library Cataloguing-in-Publication Data
A catalogue record for this book is available from the British Library

Library of Congress Cataloging-in-Publication Data
Names: Dudley, Nigel, author. | Stolton, Sue, author.
Title: Leaving space for nature : the critical role of area-based conservation /
    Nigel Dudley and Sue Stolton.
Description: Milton Park, Abingdon, Oxon ; New York, NY : Routledge,
    2020. | Series: Routledge studies in conservation and the environment |
    Includes bibliographical references and index.
Identifiers: LCCN 2019058930 (print) | LCCN 2019058931 (ebook) |
    ISBN 9780367407544 (hardback) | ISBN 9780367407537 (paperback) |
    ISBN 9780367815424 (ebook)
Subjects: LCSH: Nature conservation—Philosophy. | Naturalness
    (Environmental sciences) | Philosophy of nature.
Classification: LCC QH75 .D833 2020 (print) | LCC QH75 (ebook) |
    DDC 333.72—dc23
LC record available at https://lccn.loc.gov/2019058930
LC ebook record available at https://lccn.loc.gov/2019058931

ISBN: 978-0-367-40754-4 (hbk)
ISBN: 978-0-367-40753-7 (pbk)
ISBN: 978-0-367-81542-4 (ebk)

Typeset in Bembo
by Apex CoVantage, LLC

CONTENTS

TABLES

PREFACE

Halfway through writing this book we visited the Parque de la Papa or Potato Park, high in the Andes above the ancient Inka capital of Cusco, Peru, a place we have heard about for years. The park covers almost 10,000 hectares and includes six settlements of mainly Quechua-speaking people, the original inhabitants of the region. Apart from the highest tops of the mountains, the whole area outside the villages is farmed on an eight-year rotation (one year of crops followed by seven fallow), with a vertical stratification moving from cereals at the lowest levels, through a variety of Andean tubers, to potatoes, which in the warmer conditions being created by climate change can now be grown above 4,300 metres (Argumedo, 2008). In the steep and uncompromising conditions hand cultivation is necessary, using tools that have not changed significantly since Inka times. Thus far, the area sounds like innumerable sites all over the Andes. But in a unique twist, since 1992 the communities within the self-proclaimed boundaries of the Potato Park have been deliberately conserving the huge crop genetic diversity that the area contains: three crop wild relatives of potato, an astonishing 1,377 potato varieties and another ninety-two Andean tubers used for food. The diversity is kept through careful management of crops, plus specimens in greenhouses on the site and, in 2015, seeds of several hundred of the Potato Park varieties were taken to the Svalbard seed bank, the global repository for agrobiodiversity on a remote island offshore of Norway.

We found the visit inspirational; a heady mixture of traditional beliefs and practices with hard science. Each day's planting starts with a ritual offering of coca leaves and there is a strong emphasis on the maintenance of traditional culture, while the work was backed up with careful application of scientific knowledge and collaboration with agronomists and visiting specialists. Perhaps as important is the Quechua philosophy of three intersecting realms: *Ruma ayllu*, area of human activity; *Auqui ayllu*, realm of the sacred; and *Salqua ayllu*, community of the wild. People were deeply rooted in their own community but at the same time passionately aware of the national and global importance of what they were doing; acting locally but thinking globally. Indeed their twenty-year vision is to see a series of similar sites in other centres of crop diversity in Bhutan, Kyrgyzstan and elsewhere, and they are already collaborating with people in these places to help make it happen. But is this a protected area? It is certainly not wild in any of the usual senses of the world; the land has been farmed for centuries if not millennia. There are thousands of people living and working there. And the biodiversity being protected is mainly created by humans. We mention the Potato Park here because in the past it has been a sticking point in conservation debates, with some arguing that the fact it is protecting landraces on farmland rather than wild biodiversity in a natural ecosystem means that it could not be a "protected area" in the way that nature conservationists understand. Today opinion seems to be shifting; a huge collective effort to protect critical important genetic diversity is worth recognising as a protected area even if it doesn't fall into the traditional format of a nature reserve.

In another example, a couple of weeks earlier, and a few hundred miles west in Lima, we took part in the third Latin American and Caribbean Parks Congress, where there was a conversation about surf breaks (Scheske et al., 2019). Under Peruvian law, it is possible to protect particular lines of surf, mainly from infrastructure development. Protected surf breaks are under the jurisdiction of the navy, which means that they have protection with some muscle behind it and they are policed by Peru's enthusiastic surfing community. Some surf breaks also protect significant biodiversity, more-or-less by accident, like important seabird colonies. The reason we got into the debate is because of proposals that such areas might be recognised as "other effective area-based conservation measures" (of which, more later).

For us, both these approaches fit fairly and squarely into our vision of "area-based conservation". The world is already too altered, too threatened and conservation too under-resourced to be purist. At least, not all the time. Conservation cannot be the sole preserve of the naturalist or the conservation biologist if it is to address the colossal challenges that will continue to face ecosystems over the next few decades. A large part of what we will be looking at in the following pages is how we can bring area-based conservation out of the specialist closet and integrate it as a fundamental part of general human activities.

This book draws on over twenty years' experience of working closely with the protected area movement (Equilibrium Research, 2020), as researchers, volunteers and campaigners. In that time, we have been lucky enough to visit a representative proportion of the world's protected areas; worked with hundreds of different managers and rangers; and engaged for many years in the international policy debate, working closely with the IUCN World Commission on Protected Areas, the Convention on Biological Diversity and many governments and NGOs. Almost all the detailed examples quoted are known to us personally, although we stress that this does not imply complete knowledge of the complexities of individual sites. (Because we work mainly on land, this biases the book against marine issues, so we include some marine examples that we only know second hand.) We have also, prior to our engagement in protected areas, worked with many other forms of sustainable land management; this experience is becoming increasingly relevant as ideas about what constitutes area-based conservation constantly evolve. What follows is therefore a personal view of protected area development, and where area-based conservation may go in the future. We owe a huge debt of gratitude to the friends and colleagues we have worked, travelled and discussed these issues with over many years. They are too numerous to list here, but have our profound thanks. Remaining errors of fact of opinion are our own responsibility.

This is the third book on protected areas that we have published with Earthscan. The first (Stolton and Dudley, 1999) drew together many experts to look at different ways in which partnerships might evolve to help maintain protected areas, and the second (Stolton and Dudley, 2010) looked in detail at ecosystem services from protected areas. They both emerged during the great boom in protected area establishment when conservation professionals and others were still struggling to understand exactly what we were dealing with. This third differs both in being an authored rather than an edited volume and in attempting a broader analysis of where we are now. We are living in a significant period, when there is both a huge surge of support for conservation but also a backlash that is almost religious in its conviction that any concern for the environment is antithetical to human progress. Some of the world's current political leaders are terrifying in their disregard. For the generation of conservationists starting their careers today, there is the grim prospect of seeing many losses and setbacks; the next few decades are going to be tough, and we will need to keep our nerve. Maintaining an optimistic, long-term vision is going to be an important factor in success. We hope this short book helps contribute in some small way to achieving this.

References

Argumedo, A. (2008) 'The Potato Park, Peru: Conserving agrobiodiversity in an Andean indigenous biocultural heritage area'. In: Amend, T., Brown, J., Kothari, A., Phillips, A. and Stolton, S. (eds.) *Values of Protected Landscapes and Seascapes: 1. Protected Landscapes and Agrobiodiversity Values*. IUCN and GTZ, Kasparek Verlag, Heidelberg

Equilibrium Research. (2020) *Equilibrium Research, Practical Solutions to Conservation Challenges: 30 Years Making Waves . . . and into the Future*. Equilibrium Research, Bristol

Scheske, C., Arroyo Rodriguez, M., Buttazzoni, J.E., Strong-Cvetich, N., Gelcich, S., et al. (2019) 'Surfing and marine conservation: Exploring surf-break protection as IUCN protected area categories and other effective area-based conservation measures'. *Aquatic Conservation Marine and Freshwater Ecosystems*, vol. 29, supplement 2, pp 195–211

Stolton, S. and Dudley, N. (eds.) (1999) *Partnerships for Protection: New Strategies for Planning and Management for Protected Areas*. Earthscan, London

Stolton, S. and Dudley, N. (eds.) (2010) *Arguments for Protected Areas: Multiple Benefits for Conservation and Use*. Earthscan, London

PART I
Setting the scene

1

A VISION FOR AREA-BASED CONSERVATION

A vision for future conservation is proposed: **most of the world's lands and waters should function as diverse, resilient ecosystems.** *The roles of various forms of area-based conservation are introduced in relation to this, along with some examples of how future thinking about conservation needs to be more imaginative and less doctrinaire than in the past.*

Afon Dyfi, the river that runs down to the coast close to our house, is the traditional border between North and South Wales, in the far west of Britain. It is still just about tidal when it reaches us; indeed long ago our village of Derwenlas was a port and ship-building centre, until the railway blasted its way through and altered the course of the river. There is some fertile land for cattle and sheep grazing in the valley bottom (unsuitable for cropping because it floods), rough pasture on the hills, oak woodlands that are becoming old enough to show some natural ecological characteristics, upland moor and, further downstream, salt marsh and sand dunes. At a rough count, there are also around ten different designations of "protected area" or area of high biodiversity importance in this one rather small valley, with varying degrees of legislative power backing them up. The whole area is a UNESCO biosphere reserve, designated in 2009, and the estuary is also listed as a Ramsar site by the International Convention of Wetlands. It is recognised as an Important Bird Area by BirdLife International, probably because of a wintering population of Greenland whitefront geese, and thus in turn becomes a Key Biodiversity Area. The Dyfi forms the southern border of the Snowdonia National Park, meaning that half the valley lies inside its boundaries. A series of National Nature Reserves protects the largest raised bog in Wales, the coastal sand dune complex and much of the estuary. Further inland private bodies take over, with two reserves managed by the Royal Society for the Protection of Birds and two by the Montgomery Wildlife Trust: both these organisations charge an entrance fee, and one reserve attracts tens of thousands of visitors a year to see nesting ospreys (*Pandion haliaetus*). The coast has some level of protection as we write, under European law. At least two of the private woodlands in the valley are being managed unofficially as nature reserves. And there is a large "rewilding" project starting up.

We live in a place where the richness in biodiversity, at least in UK terms, perhaps means there has been a higher than average interest in conservation, but our situation is not really that unusual. In the last few decades, literally hundreds of thousands of "protected areas" have been set up around the world, officially and unofficially, very large and very small, and with variable amounts of planning, consultation and management input, and also with great differences in their degree of success or failure.

The growth in protected areas represents a huge surge of interest in and worry about what humans are doing to the natural world, an interest that includes ethical concerns, emotional feelings and pragmatic self-interest. For many people the progress made so far is useful but far from enough, and they call for more protection, more restoration and a fundamental rethinking of our relationship with the natural world. For others the process has already gone too far, hampering development, displacing people and placing "wildlife" interests above human interests. At a time of rapid political polarisation these debates are becoming more intense, although it would be simplistic to assume that conservation interests break down neatly along traditional political fault lines, which makes it very important at the present that we understand exactly what protected areas are, why they are there and what they might mean in the future.

The philosophy behind this book

Too many decisions about conservation – too many decisions about anything – are made on the basis of ingrained prejudices, peer pressure, lazy thinking or on simply doing what people have done before. In a rapidly changing and divided world, there is too much at stake to drift into a conservation policy without knowing exactly where we are going. When we talk about protected areas and other forms of area-based conservation we need to know as clearly as possible "why" they are there, "what" they are for, "where" they should be and "how" they are to be established and managed. In this preparatory section we try to outline where we see the priorities and end-goals of managing the planet's natural resources and where area-based conservation fits into this strategy.

The vision driving the arguments in this book is simple: that in the future "most of the world's lands and waters should function as diverse, resilient ecosystems". It is a vision that draws on, acknowledges and offers a way to fulfil the Convention on Biological Diversity's (CBD) goal of "living in harmony with nature" which states that "by 2050, biodiversity is valued, conserved, restored and wisely used, maintaining ecosystem services, sustaining a healthy planet and delivering benefits essential for all people" (CBD, 2010).

For the past twenty years there have been debates about percentage targets for protected areas: we'll discuss these more (see Chapter 6), but for now note that countries have to date collectively protected around 15 per cent of land and 7.8 per cent of marine area, according to statistics on the Protected Planet website at the time of writing. The overwhelming evidence suggests that to maintain many of the world's critical life-support systems we need to be way more ambitious, summarised in "most"; maintaining and restoring natural ecosystems is not something to be isolated as an issue of minority concern but as a central plank of human progress and we thus need to devote the majority of the world to some form of area-based conservation. This need not be quite as radical as it sounds, and it builds on discussions and debates that have involved many people over the last few years; we are not claiming this is a particular original perspective. There are multiple reasons for aiming so high, stretching from the strictly pragmatic to the cultural and emotional, and these will be looked at in detail later. The phrase "lands and waters" makes the point that we need to be concerned about more than just terrestrial and freshwater ecosystems but, perhaps to an even greater extent, conservation effort needs to focus on the world's coastal zones, seas and oceans. "Diverse, resilient ecosystems" implies ecosystems capable of functioning in the long term and supporting as much as possible of the world's existing biodiversity and ecosystem services in the face of a rapidly accelerating suite of environmental changes. This means that we are talking about ecosystems that are not dramatically simplified, as is the case with plantation agriculture, monoculture tree plantations, intensive, modified grazing or aquaculture sites: these may all have their place within a landscape or seascape but should be a minority land and water use on a global scale and subject to important constraints.

What we do not say is "natural ecosystems". Today we know that all ecosystems have been modified to some extent or other by human activity, including those – like the Amazon, Antarctica and the vast plains of southern Africa and North America – that many people still regard instinctively as "wild". The degree and speed of modification is increasing enormously as a result of climate change, land use change, pollution, resource extraction, species extinctions and the impacts of invasive species. Unfortunately, we have barely seen the beginning of this process.

Some people argue that this means that the whole concept of naturalness is no longer relevant, or is much less relevant (e.g., Low, 2002; Pearce, 2015). We disagree and have argued against this previously. Ecosystems change and our own actions have made many alter far more rapidly than would be expected in other circumstances, but ecosystems also adapt and reach a new balance. We have suggested the term "authenticity" may be a useful alternative to "naturalness" in recognising these human-modified but still diverse and resilient ecosystems. We define an authentic ecosystem as "a resilient ecosystem with the level of biodiversity and range of ecological interactions that would be predicted as a result of a combination of historic, geographic and climatic conditions in a particular landscape" (Dudley, 2011). In other words, the general principles of naturalness remain important, but there is little point in getting too hung up on trying to retain or recreate some mythical "pure ecosystem" untouched by human hands, because such places no longer exist.

The vision we outline earlier is necessarily long term; it will take decades, probably centuries to achieve. Furthermore, we expect to see some major reversals along the way; this book starts from a position of short-term pessimism but long-term optimism. The extent to which humans can be "observers" of future ecosystems or active "stewards" will also depend on many things, including social and political attitudes of future generations. Attaining our vision certainly implies a good deal of stewardship – active intervention, management and perhaps above all restoration – in the medium term. It means ensuring concepts like ecosystem-based adaption become the norm rather than the exception as the world learns to cope with the stresses that we have created and looks to natural solutions to solve them. It also implies developing, or perhaps rediscovering, a more respectful attitude to the rest of nature and the way that it functions.

The importance of area-based conservation

"Protected areas" is a collective term for national parks, nature reserves, wilderness areas and the like and has official definitions and standing (see Chapter 4). But today there are a growing number of recognised management approaches that result in conservation but fall outside the definition of a "protected area", and these are likely to become increasingly important. In this book we will use the term "area-based conservation" to describe the gamut of approaches that are used or are developing.

Indeed, if this book has a central message it is in tracing out the development, or more properly the expansion, "from protected areas to area-based conservation". This does not mean that protected areas are redundant; indeed we will argue that they are essential and we still need many more. But they are no longer the only tool for protecting biodiversity and other ecosystem services in specific sites. This in turn implies another associated development, "from management to stewardship". Again, this is far from implying that management is no longer needed; in many cases we need more and better management, but this needs to be embedded within a wider social engagement in and responsibility for land and water, with many different stakeholders involved.

We see area-based conservation playing the critical role in stabilising and recovering the life-support systems for the planet: hence this book. In our own professional lives, we have moved from working on various forms of sustainable management – forests, organic agriculture and pollution reduction – to focusing on protected areas precisely because we think these offer the most concrete way of addressing ecosystem degradation and loss for the foreseeable future. But the scale of what we believe is necessary means that the world must embrace a much wider suite of responses than what we now know as protected areas. We are not advocating a future where huge areas of land are set aside to conserve "nature" for the rich minority who can afford the price of an entry fee. Nor are we assuming that the responsibility for conservation should rest disproportionately in the hands of countries that still retain large areas of near-natural ecosystems. For many reasons, this needs to be a global effort, almost certainly the most ambitious and defining global project of the twenty-first century.

Today we have some but by no means all of the necessary elements in place, and those that exist are far from completely implemented. Commitments to conservation on this scale imply that the large majority of people on the planet accept these steps as necessary, which in turn requires whole new ways of approaching the negotiation, governance and management of the planet's resources. We are by no means there yet, but things are moving fast.

Changes in attitudes and understanding over the last decade mean that we are on the edge of some exciting new developments in area-based conservation that will revolutionise the way we approach natural resource management. These new approaches give us much cause for optimism but are certainly not risk-free, and their strengths and weaknesses will be unpicked later. Some of the assumptions currently being made about how societies will react and wish to manage land and water that they control are naïve. To some extent the divisions between factions in the conservation debate are wider now than they were a decade ago. There is an urgent need for a greater level of consensus, more compromise and a wider circle of advocates.

The rest of the book looks at how attaining a situation where "most of the world's lands and waters function as diverse, resilient ecosystems" might be achieved. It looks at what has been done so far – at the status of area-based conservation worldwide, its strengths and weaknesses. It then discusses why this is not enough, looking both at the level of threats to ecosystem stability and at the weaknesses still to be addressed within protected areas and their relatives, including current attempts to address these. Finally, we suggest where we need to go from here. This includes both long-term proposals for managing the world's authentic ecosystems and some short- and medium-term tactics that we believe may be necessary to address critical emerging threats.

References

CBD. (2010) *Strategic Plan for Biodiversity 2011–2020 and the Aichi Targets*. Secretariat of the Convention on Biological Diversity, Montreal

Dudley, N. (2011) *Authenticity in Nature: Making Choices About the Naturalness of Ecosystems*. Earthscan, London

Low, T. (2002) *The New Nature*. Viking, Camberwell, VIC

Pearce, F. (2015) *The New Wild: Why Invasive Species Will Be Nature's Salvation*. Icon Books, London

2

WHAT ARE WE AIMING FOR?

Having once decided on an overall vision, a "what" we are aiming for in terms of area-based conservation, it is important to consider more clearly why this is needed, where it needs to take place and how area-based conservation might be managed and governed. Some initial thoughts are given in the following. Finally, we look in a little more detail at the question of whether or not naturalness retains any relevance in a rapidly changing planet.

After the 2016 World Conservation Congress in Honolulu, Hawaii, we went to Kauai, one of the outlying islands. We walked up high into the mountains, where rainforest stretched down to the sea over rugged cliffs and there were high Alpine wetlands. The area looked like untouched nature, but the reality is very different. Many of the plants we were seeing are invasive; since humans arrived ninety-five endemic bird species have become extinct, with several more probably extinct and many other threatened. Most of the birds we did see come originally from other places. The "wilderness" stretched out beneath our feet is almost entirely the result of human activity. What does conservation mean in a place like Kauai? Does it have any real meaning? Do we just give up? There is still a tendency amongst conservationists to say that once an ecosystem has been damaged it has lost all value. We've often been told that Cameroon has virtually "no forest left", by which the speaker means no forest that has not been logged over and the largest trees removed. That may be true, but again does that mean we forget about Cameroon? In the following chapter we unpick the conservation vision a little, look at the steps needed to reach it and at the relevance of such strategies in a world that has already been profoundly transformed.

If the "what" in the title of this chapter is a healthy planet, where "most of the world's lands and waters should function as diverse, resilient ecosystems" we need to start looking at the "why", the "where" and the "how".

Although it is generally considered a bad form of debating, maybe we could start with what we are "not" aiming for. There is an old debate about "sharing or sparing": about whether to promote landscapes and seascapes in which nature and human activities exist in the same space, or in which they are rigorously separated between protected areas and production areas. We are not suggesting a black-and-white split between nature and not-nature, or promoting a world in which beautiful, untouched nature reserves are contrasted by intensive farmland and forest plantations, powered with agrochemicals and using the latest in genetically modified technology. This doesn't work; there is no impermeable barrier between nature and farmland for instance. Some of our earliest research work looked at spray drift of pesticides and the many routes that nitrate fertilizers can take away from their target crops and into the wider environment (Dudley, 1985). The quantities being used today are almost literally an order of magnitude greater than when we were first looking at the problems. Suffice to say here that if we don't get management right in production landscapes and seascapes the impacts will spill over into the rest of the environment, often huge distances away from the source of any particular contamination. But these strict divisions between nature and culture have other implications. They mean, for example, that most people would be forced to live in contaminated, impoverished environments, interacting with "nature" through a television, tablet or smartphone, or during occasional holidays for those who could afford them.

Why conserve?

A few years ago, we were doing a road trip in Viet Nam, heading north out of Hanoi for several hours to reach Van Lonh nature reserve. On the way we saw just four birds, all egrets, which presumably don't taste good. Viet Nam is one of the most heavily hunted countries that we know. Curious about what the difference would be in the UK we made a similar count driving between Wales and Bristol, a journey of around three hours through the mountains and valleys. We counted well over a thousand birds out of the car window. There wasn't a huge variety in our bird list: we missed most of the smaller birds in the hedgerows, but there were a lot of corvids, quite a few birds of prey, odd fields of gulls, blackbirds skimming away over fields, an odd heron spotted by a river.

Efforts at species conservation tend to focus, not surprisingly, on extinction threats. We hear constantly that we are in the midst of an extinction crisis and the IUCN Red List provides a depressing list of statistics of loss. This is critically important, and we are certainly not arguing otherwise. But a change, usually but not invariably a loss, of abundance can be almost as important, and can eventually spiral on downwards to threats of extinction. The populations of birds, mammals and particularly ocean fish species that we see today are often a fraction of what they were in the not too distant past; conversely populations of invasive species have exploded, bringing with them an inherent instability and the possibility of a spectacular crash. It is frightening how we can sit on warm summer evenings in many cities, with the lights on and the windows open, and scarcely attract a moth or a flying insect. It is disheartening to wake up on spring morning in temperate countries and hear nothing of the famous dawn chorus of birds singing their hearts out as they try to defend territories or attract a mate. Or walk through pastures made up of a single grass species, forest stands without undergrowth, tidal rockpools with nothing but a handful of the hardiest sea anemones and gastropods hanging on . . . we could continue. These are all changes we've seen during our lifetimes, and they are symptomatic of a far wider loss.

The reasons for area-based conservation are not only about protecting the rarest species, or a representative sample of all species, important though these issues are within conservation strategies. They will be getting a fair amount of attention in the following chapters. But area-based conservation is also about preserving healthy populations of commoner species, living together in functioning ecosystems. These kinds of values tend to be missed in planning tools that focus only on the rarest or most "important" plants and animals, or the most intact forests. We need both to fulfil our vision of functioning, diverse and resilient ecosystems.

Added to this are the services that healthy, functioning habitats and ecosystems provide to humanity: water and food security, human security through disaster risk reduction and the mitigation of climate change, medicines, materials, space to exercise and cultural and spiritual benefits from being able to access beautiful, peaceful places. And finally, and unexpectedly, some protected areas in particular are being identified, lobbied for and designated in large part to defend the embattled human societies that live there.

We therefore can divide the reasons for area-based conservation into three:

- **Diversity**: maintaining a full natural complement of species on the planet, interacting together in functional ecosystems;
- **Abundance**: ensuring that the other species we share the planet with are not reduced to fragile, rump populations, barely able to survive without continuing human intervention; and
- **Human benefits**: the ecosystem services derived from natural and semi-natural ecosystems, discussed in Chapter 9, and as a way to protect threatened human cultures.

These values cannot wholly be addressed in a series of isolated protected areas, however well-managed or resourced they happen to be (and remember that many are woefully under-resourced and poorly managed in consequence). They need big spaces, a big vision. But that also has some implications for the way that such a vision might be achieved on a crowded planet.

Where to conserve?

We will, later in the book, summarise some of the ways in which conservation priorities are set. Planning tools and priority setting exercises are extremely valuable.

But in the wider picture that we sketched out earlier they are only part of the picture. An area-based conservation policy that *only* focused on the most valuable places for biodiversity would miss out many places that are critically important for ecosystem services, or that stand a chance of supporting commoner species and ecosystems at levels that come close to or reach the original abundance of life that has so often been pauperised by our own mismanagement and overuse.

Harvey Locke and colleagues have proposed that the land surface of the planet can be usefully divided into three conditions: named in shorthand "cities and farms", "shared landscapes" and "large wild areas". Cities and farms cover 18 per cent of the land, contain 75 per cent of the population and currently have 6 per cent in protected areas. Shared lands cover 56 per cent of the land, contain 25 per cent of the population and currently have 14 per cent in protected areas. Large wild areas cover 26 per cent of the land, contain 0.1 per cent of the population and have 24 per cent in protected areas (Locke et al., 2019). Scientists can and will quibble endlessly about the proportion of land in different areas and about the relative integrity of different ecosystems.

But from our perspective here the concept is interesting mainly because it highlights the different objectives and the diversity of approaches that will be needed within area-based conservation, to fulfil an overall vision of a healthy, sustainable planet. A lot of the most threatened species and ecosystems will be in the shared lands, such as places where natural tropical forests are being squeezed out by oil palm and soy plantations, or where mixed grazing takes place on relatively natural grassland and savannah. Some fairly strict reserves will be needed here to resist the enormous pressures facing many such landscapes, but we will also need to be working with protected areas that are based around cultural landscapes, or with places outside protected area systems where changes in management can result in a boost for nature. Protecting intact ecosystems will also be critical but will require different strategies. A proportion of the wildest areas will be relatively low in species – such as the vast boreal – but critical for ecosystem services. On the other hand, recent research suggests that wilderness areas play a disproportionate role in reducing extinction risk for terrestrial biodiversity (Di Marco et al., 2019). Cities will have completely different needs and opportunities for area-based conservation than more natural areas and so on.

A similar analysis has not yet been carried out for the ocean, but the same three conditions could well be applied with some slight wording changes. There are port facilities, mining sites, heavily used coastal areas, aquaculture sites and places where fishing is so intense that the whole structure of the ocean has changed – for example through sea-bed trawling – these could be termed something like "settlement and intensive fishing". Then there would be "shared seascapes", less heavily used areas where small-scale fishing takes place alongside relatively natural ecosystems, areas of open ocean where fishing is occasional, offshore wind farms that contain important marine life around them and so on. And finally, as on land, "large wild areas" where there is little or no fishing, mainly oceanic areas and some remote offshore islands and atolls. Perhaps even more so than on land, such an analysis would need to take account of contamination of various kinds: the dead zones where eutrophication has knocked out all life, reefs being destroyed by invasive starfish and, increasingly, the impacts of ocean acidification and plastic pollution. Developing a more fully thought out set of conditions and their relative areas would be a useful contribution to marine conservation planning.

Deciding where to conserve is therefore complex. While conservation planners like to juggle with algorithms and come up with utopian schemes, the reality in most places is that choices are and will continue to be heavily influenced by politics and business interests at both a national and local scale and conservation interests often have to take what is offered rather than what they would, in an ideal world, aspire to. And climate change brings a massive new layer of complication; what might be an ideal conservation area today might not be quite so attractive in a few years' time.

How to conserve?

The "where" to conserve is therefore intimately bound up with "how" to conserve. And one of the more exciting recent developments that we will be exploring in the following chapters is that the "how" has literally exploded into a plethora of new opportunities over the last few years, which the conservation movement as a whole is still struggling to assimilate. There are dangers here; we could see lots of greenwashing as people and governments claim dubious conservation gains without making any real changes to what they are doing. But at the same time there are also some very important new opportunities that we are only just beginning to explore.

So, in addition to protected areas we have a range of other approaches that sit under the banner of area-based conservation, or perhaps it might be more accurate to say are just starting to be accepted at the table. Community conservation areas, indigenous protected areas, territories for life, privately protected areas, the new "other effective areas-based conservation measures", areas of connectivity conservation, sustainable forest management and so on.

We will be exploring these in detail later. But for now it is important to note that the changes affect both what happens within a site of area-based conservation – in other words the "management strategies", the extent to which people are or are not recognised as part of the ecosystem and so on – and also the "governance strategies"; who is in charge and who makes the decisions.

Does "natural" matter anymore?

We have already mentioned ideas about "authenticity" and our earlier attempts to make sense of what "natural" means in a planet increasingly shaped by humans (Dudley, 2011). Most of the planet – in fact virtually the whole planet except for parts of Africa – has already been dramatically and irreversibly altered before written history, by the extinction of many top predators, large herbivores and others. The mammoth, sabre tooth tiger, the wild horse species of South America, many of the giant flightless birds are all gone, leaving an inherently unstable set of eco-systems behind; in evolutionary terms these changes happened a couple of minutes ago and things have not adjusted. Then the huge changes to vegetation created by the Neolithic Revolutions, which occurred separately in Eurasia and the Americas, the continued land use changes and pollution load, and what Jeff McNeely (2001) calls the "great reshuffling" of species moved by human agency into ecosystems for which they are not adjusted. Add climate change to the mix and there is not that much completely natural to hold onto. When faced with this scale of change some scientists shake their heads and say we should let nature find its own balance.

We disagree, albeit listening carefully to these voices. Life is persistent and eco-systems do adapt and function, even if they are not quite as they once were; the driving force of nature is change. There is good evidence that diverse, function-ing ecosystems provide more security in terms of ecosystem service delivery than fragmented and impoverished ecosystems (Oliver et al., 2015). But we need to give nature a fighting chance. There are places where currently ecosystems remain rich, relatively natural at least in the context of the last few thousand years, in balance and culturally and biologically important. Many of us believe it is worth making a substantial effort to maintain such places into the future, even under climate change. In other parts of the world, where ecosystems are likely to change even more dramatically, we may indeed just need to stand back and see what happens or intervene more fundamentally to move or protect species that no longer have living conditions in the place they used to call home. Concepts like "novel ecosys-tems" have developed to describe and to prepare ourselves for the scale of changes that may take place (Hobbs et al., 2009). We return to the more apocalyptic visions of the future and what they might mean for conservation in the final part of the book.

What this all means is that "natural" or "authentic" ecosystems do indeed have value, but these values are not absolute, ecosystems are not fixed and in many cases we are going to have to make up conservation as we go along. Furthermore, our understanding about the meaning and significance of naturalness has shifted dra-matically over time, from humans existing as an unthinking part of nature, to a gradual and then a conscious separation from and rejection of the natural and now a gradual, incomplete and hesitant re-engagement. Attitudes to naturalness are not value-free and attract strong levels of emotion. Whenever we attempt to manage ecosystems, we are making choices, often unconsciously, about levels and types of authenticity: in most cases there will not be a single "authentic" ecosystem but a large number of alternatives.

References

Di Marco, M., Ferrier, S., Harwood, T.D., Hoskins, A.J. and Watson, J.E.M. (2019) 'Wil-derness areas halve the extinction risk of terrestrial biodiversity'. *Nature*. Doi:10.1038/s41586-019-1567-7

Dudley, N. (1985) 'Environmental and economic constraints on spraying systems'. *1985 Brit-ish Crop Protection Conference – Weeds*, pp 1135–1143

Dudley, N. (2011) *Authenticity in Nature: Making Choices about the Naturalness of Ecosystems*, Earthscan, London

Hobbs, R.J., Higgs, E. and Harris, J.A. (2009) 'Novel ecosystems: Implications for conservation and restoration'. *Trends in Ecology and Evolution*, vol. 24, no. 11, pp 599–605

Locke, H., Ellis, E.C., Venter, O., Schuster, R., Ma, K., et al. (2019) 'Three global conditions for biodiversity conservation and sustainable use: An implementation framework'. *National Science Review*. Doi:10.1093/nsr/nwz136

McNeely, J.A. (ed.) (2001) *The Great Reshuffling: Human Dimensions of Alien Invasive Species*. IUCN, Gland, Switzerland and Cambridge

Oliver, T.H., Isaac, N.J.B., August, T.A., Woodcock, B.A., Roy, D.B., et al. (2015) 'Declining resilience of ecosystem functions under biodiversity loss'. *Nature Communications*, vol. 6, no. 10122. Doi:10.1038/ncomms10122

3

A BRIEF HISTORY OF THE MODERN PROTECTED AREA MOVEMENT

The creation of modern protected areas is the largest and fastest conscious change of land and water management in history, already covering 15 per cent of the planet's land surface and a growing area of coastal waters and ocean. They first emerged in a small way at the end of the nineteenth century, mainly protecting sites for particular iconic animals or extraordinary scenery. Through the first half of the twentieth century there was little further change, but then from the 1970s onwards an extraordinary boom in creation occurred; most of the world's protected areas will have been created during the lifetimes of most readers of this book. But they remain contested spaces, far from universally popular, and are neither monolithic in management and control, nor necessarily permanent. Opinions about what constitutes a protected area are also changing fast, and a number of other area-based conservation approaches are emerging, which will deeply influence the way in which we manage natural ecosystems. We are approaching a critical juncture in the future of global conservation. An understanding of the range of opportunities available for area-based conservation, their strengths and weaknesses and the opportunities they present, is therefore particularly timely.

In a satellite image, the Brahmaputra River stretches like a broad ribbon across north-east India and into Bangladesh, rising in the Himalayan foothills of Arunachal Pradesh and making its way gradually west and south to the Bay of Bengal. Not so much a single river as a latticework of separate channels, the banks crumble and shift with the seasons and storms, moving by a mile or more in any single year. It was along the southern bank of the Brahmaputra that the Kaziranga Forest Reserve was established in 1902, a vast area of elephant grass, marsh and dense tropical forest that eventually became Kaziranga National Park and later was also recognised as a UNESCO natural World Heritage site (Mathur et al., 2007). Today it supports the world's densest populations of both one-horned rhinoceros (*Rhinoceros unicornis*) and tigers (*Panthera tigris*) (Balmford, 2012).

As a protected area, Kaziranga demonstrates many of the complexities that we will be discussing in this book. It was set up as a result of lobbying by Lady Curzon, wife of the Viceroy of India (Choudhury, 2004). Far from a conventional member of the British upper classes, Mary Curzon was an American from a wealthy business family who was persuaded of the need for a protected area by Balaram Hazarika, a noted Assamese tracker. A more precise example of colonially imposed conservation is hard to imagine, and this narrative attracts criticism from many Indian observers who question both the conservation motives and the importance of Curzon's role (Saikia, 2009). And Kaziranga is also in many ways a quintessential example of so-called fortress conservation, as rangers seek to defend wildlife populations against poachers who are drawn constantly to what is now almost certainly the world's richest pickings of rhino horn and tiger bones, both ridiculously valuable on the black market. When we visited around the turn of the century there were about 250 guard posts throughout the park, each supplied with a shrine and a cat, backed up by a huge network of spies and informers. Today their work is further assisted by drones and other forms of electronic surveillance. And in the politically troubled environment of north-eastern India tigers are protected through force of arms; there is little mercy on either side, with shootouts between rangers and poachers.

It has become fashionable to decry this kind of conservation as anachronistic: the vestiges of colonial thinking in which animals are preserved in ruthless ways so that the rich and privileged can take a vacation and see them, with little thought to the local people who might be inconvenienced in the process. "Inconvenience" can here include dispossession of traditional lands and resources (Duffy, 2010) and injury or death from top predators. We will be looking at many of these issues, and the critics have important points to make. And yet . . . we almost certainly would no longer have one-horned rhino walking the planet if it wasn't for Kaziranga and a small number of other effective protected areas that have kept the species alive, and the sacrifices of dedicated conservation staff who often risk physical persecution and personal dangers for scant personal rewards (Kunwar, 2009). In fact, we would probably have no rhinos left alive if it were not for protected areas.

Conservation often takes place in tough conditions and is frequently messy, imperfect and compromised; some of the complexities are discussed later. The interplay between human rights, the rights of other plant and animal species, the need for ecosystem services, trade-offs between economic and social development, and the imperative of maintaining a viable, living planet do not make for simplistic answers. In this book we will be arguing that one critical response is to secure large areas of more-or-less natural ecosystems, and that various forms of area-based conservation are at the present time the most effective way of doing so. But we are certainly not pretending that this is a simple or straightforward option.

Early "protected areas"

Humans have set aside natural, or nearly natural, areas of land and water for centuries, probably millennia, and for a wide variety of reasons. Reserving important forests for their timber supply has been practised throughout the world. The Romans enacted laws to protect the famous cedars of Lebanon in 134 AD, when Emperor Hadrian ordered boundary stones to be placed around remaining forest and designated them "Imperial Domain" as a way of ensuring that a valuable timber supply for ship building was not exhausted (Abu-Izzeddin, 2013). In the Arabian Peninsula, conservation areas called "hima" were created to protect grazing land and other ecosystem services, probably in part as a move in the ancient conflict between mobile pastoralists and settled farmers (Bagader et al.,1994). Eliminated by the House of Saud, they are now making a gradual re-emergence. In countries with steep mountain slopes and risk of avalanches and landslides, like Japan and Switzerland, forest protection stretches back hundreds of years as a safety measure for valley-bottom settlements (McShane and McShane-Caluzi, 1997). Many people identify particular places – groves, lakes, waterfalls and mountains – as "sacred natural sites" and accord them strict protection; thousands still exist today around the world (Verschuuren et al., 2010). Pacific island communities have for centuries set aside areas of coastal waters on a temporary or permanent basis to rebuild or maintain the fish stocks they rely on to survive (Govan et al., 2008). And wherever the rich were powerful enough to impose their will, areas of land were conserved as their private hunting preserves: the English word "forest" originally meant a tree-covered area set aside for royal hunting and under the protection of the king (Schama, 1995). The concept remains today, particularly in Africa, with hunting reserves set aside for the wealthy to shoot large animals like lions and elephants; the relationship of such areas with conservation strategies is controversial and will be discussed later. Few if any of these areas were established for reasons that we would now recognise as "nature conservation" in any very formal sense, but all were set up by people who recognised that it is important to retain natural ecosystems for the values that they maintain. This recognition should be the foundation of area-based conservation today; but other motivations have slipped in and perhaps in some places we are not so much leaving space for nature as using nature as just another commodity.

The modern protected area movement

Many of these "early" protected areas remain under some form or another of protection today, and we will be discussing them and their relationship with contemporary conservation concerns. But the main focus of our attention in this chapter is on the modern protected area movement, which for many historians began when US President Ulysses S. Grant signed Yellowstone National Park into law in 1872, creating what is widely claimed to be the world's first "modern" protected area. In fact, Yosemite in California had been set aside for the public good over ten years previously, creating the precedent for Yellowstone, but Yosemite only became a national park after President Roosevelt camped there in 1903 (Sheail, 2010). And royal parks in London, Berlin, Paris and Prague had already opened their gates to the public in the early 1800s, suggesting the American model was more a continuation of a process rather than a unique innovation (Jones, 2012). Significantly, much of the push to create both the first two US national parks came from fears that the areas, only relatively recently "discovered" by European settlers, might face the same fate as Niagara Falls. An early victim of commercialisation, Niagara Falls had become seen as an area where much of the natural beauty had already been spoiled by uncontrolled building and tourist development (Carruthers, 2012).

Other protected areas soon began to be established, but slowly: a few in the United States, places like the Blue Mountains National Park outside Sydney, Australia, and a handful of game reserves in the European colonies in Africa and India. At the same time, smaller nature reserves started to appear as people recognised threats to individual species and habitats and purchased or set aside land for their protection. In the UK, the National Trust obtained its first reserve, five acres above Barmouth in North Wales, in 1895; the Trust is now the UK's largest private landowner. Simultaneously, parallel steps were taken to conserve other natural resources such as the forest reserves established in many of the European colonies, which were originally identified simply as timber resources but a proportion of which were later elevated to the rank of full protected areas (Howard et al., 2000).

But all these areas were insignificant on a global scale; a trickle of new sites in the early years of the century and long periods of virtual inaction during two World Wars and decades of political and social upheaval. The UK did not establish its first national park until 1947, some seventy-five years after the designation of Yellowstone (McEwan and McEwan, 1982), and the UK was one of the first countries in Europe to take such a step. The concept of protected areas remained virtually unknown and irrelevant in those countries still containing huge functioning ecosystems, such as the vast forests of the Amazon or Congo Basin. Marine protected areas were virtually unknown. When the International Union for Conservation of Nature (IUCN) first started talking about a goal of 10 per cent of the land surface under protection, sometime in the 1970s, it seemed a Utopian dream, wholly divorced from reality.

Then, soon after 1970, something happened. The 1972 Stockholm Environment Conference was largely recognised as kick-starting the modern environmental movement, bringing a younger, angrier and more politicised group of people into areas formerly the preserve of old-style nature conservationists (who until quite recently thought nothing of obtaining a "specimen" of a rare bird with a charge of buckshot). The *Ecologist* journal published *Blueprint for Survival* (Anon., 1972) and the Club of Rome released *Limits to Growth* (Meadows et al., 1972), pointing out the risks of natural resource depletion. Global protected area coverage was then little over 2 million km^2; forty years later it was virtually an order of magnitude greater and still accelerating (Bertzky et al., 2012).

The change is almost certainly the fastest conscious change of land and water management at such a scale in the planet's history. It reflects also a profound change in attitudes towards natural resources, biodiversity and our role as members of and increasingly also stewards of the natural world. Since the millennium, there has been a corresponding boom in establishing marine protected areas in coastal waters and – really just beginning – in the high seas.

Opposition to protected areas

While research shows that protected areas generally receive high levels of public support (Kümpel, 2014), these moves have not been universally welcomed. Many business interests have protested long and hard about "locking up" natural resources and thus slowing rates of economic growth, job potentials and development (Dudley et al., 2015). Others refuse to accept the notion that such areas are off limits at all. Shell's *2013 Sustainability Report* puts the matter bluntly: "Shell aims to operate responsibly and transparently in protected areas that are rich in biodiversity, often working in collaboration with environmental experts" (Royal Dutch Shell, 2013). More intractably, the ways in which protected areas have in some cases been established have created serious human rights concerns; traditional communities being forcibly relocated outside a park boundary and/or denied access to traditional resources such as wild food, timber and water. For some human rights activists and organisations defending the rights of indigenous peoples, protected areas are the wrong model for conservation (Duffy, 2010) and should be drastically reduced, changed or even eliminated altogether. This view is gathering momentum in some areas, and we need to take it seriously. It is also an issue where emotions sometimes overrun evidence, with unrealistic claims made about both the importance of conserving "wilderness" areas without people and conversely about the ability for rapidly growing, complex societies to manage natural resources in the absence of enforced controls. We'll return to this in detail later in the book but for now note that top-down, authoritarian conservation – almost always associated with governments that are top-down and authoritarian in many other ways as well – creates a mass of problems that not only undermine human rights but eventually undermine conservation as well. This is not to deny that there will sometimes be hard choices to make between needs of local human communities, the needs of other species and the needs of a larger national or global community. But every situation is unique, it is often incredibly difficult for outsiders to make judgements about what is "right" or "wrong" and most negotiations end up in a compromise.

Hard choices in central Viet Nam

A few years back I was sitting with a group of villagers on the boundary of a national park in the Truong Son Mountains of central Viet Nam. Despite the designation of the protected area, the villagers had continued hunting, having no other obvious source of meat. Now a World Bank-funded project was increasing ranger capacity and patrolling efficiency, thus reducing opportunities for poaching and effectively cutting off this particular supply of food. Plans for access and benefit sharing agreements that were meant to find ways to make up the shortfall had not yet emerged from the government and local people were caught in a trap; more effective conservation was likely to deprive them of an important source of protein and other nutrition, and protected area designation had made their life more difficult. But it was less difficult than the alternative. On the other side of the mountain, the forest had been designated to be cleared for a pulp plantation, and the people's traditional lands had been lost altogether; there were no access and benefits sharing agreements, delayed or otherwise, and no earnest groups of people trying to sort the problem out. And food options were quickly closing off anyway as larger mammals had already been hunted out. Looking at the protected area in isolation would suggest that people were far worse off as a result of the national park; looking at the wider picture and what was likely to have happened otherwise tells a slightly different story. **ND**

Another example from further north in Viet Nam, close to the Chinese border, shows that changes caused by conservation actions can have unexpected results, not invariably negative. A group of us were talking with a local ethnic minority community about a ban on hunting a particular species of monkey that was threatened with extinction. As conservationists we were relieved by the ban; a highly intelligent primate was right on the edge and urgently needed protection. But we were also concerned about the social impacts: would bored men start taking more opium, would the ban undermine traditional cultures, would frustrations be taken on women and children? The men's replies were bland and noncommittal. But when it was the women's turn to talk things suddenly changed. It turns out the women were delighted by a hunting ban; monkeys weren't good eating, the men just sat around in the forest getting drunk and best of all, without hunting as an excuse their husbands had been forced to help out with the farming. "My husband even helped clean out the pigs!" said one woman amongst peals of laughter. The men started to relax and laugh as well. This wasn't the result we were expecting. It did indeed suggest that the ban had created a societal change, but one we felt more comfortable about.

These two examples from Viet Nam are not included to show that conservationists always get things right – of course we don't. But things are seldom clear cut. Negotiating conservation approaches that balance the needs of different rightsholders and stakeholders is one of the critical challenges for area-based conservation over the coming decades, and simplistic answers are seldom the right answers. Nor is it always possible to guess what the results will be, meaning that all conservation projects need to be adaptive, looking at what is happening and making changes if things seem to be going wrong.

Changing motivations for protection

At the same time, the reasons for establishing protected areas have evolved, and continue to evolve, widening the pool of people taking part in conservation. Lady Curzon was interested in saving the rhino; it was a concern that was mirrored by moves in North America to set aside land for the fragmentary remains of the continent's once-vast bison herds (Small, 2016), and early protection for iconic wildlife in sub-Saharan Africa (Fitter and Scott, 1978). These parks almost certainly saved species from going extinct in the wild. But many of the first national parks in the

United States were more about preserving wild scenery and landscapes, driven by writings of visionaries like John Muir: "Thousands of tired, nerve-shaken, over-civilized people are beginning to find out that going to the mountains is going home; that wildness is a necessity" (Muir, 1901). The earliest national parks in the United Kingdom were principally aimed at giving working people access to the countryside for recreation, in the face of resistance from wealthy landowners. Britain's first national park in the Peak District, an area of upland moors in the industrial heartland of England, was designated by a Labour government following pitched battles on the moors as hikers (led by the Communist Party) fought their way across against massed ranks of gamekeepers armed with pikes and sticks (Hey, 2011).

Since then motivations have continued to change and develop. The word "biodiversity" did not appear until the mid-1980s, but once recognised, the concept broadened conservation aims from being focused on species or groups of species towards the maintenance of ecosystems containing myriads of species, in many cases still not described by science. Tourism has also become a major driver. As we shall discuss later, wildlife tourism is the largest source of foreign exchange for a number of countries. Significantly, it often brings money to rural areas with few other economic options and its importance is not confined to poor or developing countries. More recently, protected areas are being established for the ecosystem services that they provide, including water, disaster risk reduction (such as the use of natural vegetation to ameliorate landslides and floods), climate change mitigation or to protect economically important crop species and other aspects of food security (Stolton and Dudley, 2010). Marine and freshwater protected areas are set up to help protect fish species of economic or subsistence value to local communities. Many natural sites also have sacred values to local communities as well, and if these are included in protected areas they need to be reflected in management.

Generally, such changes in the reasons for protection are not "instead of" but "as well as" earlier objectives, making managers' lives increasingly complicated as one area of land or water is expected to supply multiple (and changing) benefits (Watson et al., 2014). The Blue Mountains National Park outside Sydney, Australia, was originally set aside primarily as a recreational resource, easily accessible by rail or road from the New South Wales capital and with a mass of walking tracks, tourist towns and viewing sites from high cliffs at the park's edge. But the Blue Mountains also support important biodiversity, protect water resources for Sydney including in Lake Burragorang (a reservoir), form an important carbon store and so on. None of these were recognised when the park was established, but today tourism needs have to be balanced with these other, complementary, values (Reid, 2013).

Few if any of the world's larger protected areas are without numerous, overlapping and sometimes mutually exclusive values and these may also change over time: the significance of natural ecosystems in mitigating and adapting to climate change has only been widely recognised in the last fifteen years or so but is now a major concern in protected area policies. Other values are likely to emerge in the future. Professionals with expertise in wildlife may be expected to handle millions of visitors, manage water resources, understand the complexities of carbon sequestration and deal sensitively with communities who regard part of or the entire park as sacred. In some other protected areas management is focused primarily on objectives such as tourism, with conservation expertise either completely lacking or only rarely available. The rapid rate of expansion, along with an equally rapid development of concepts about conservation, have sometimes run ahead of the ability of those managing protected areas to keep up.

At the same time, management and governance structures are changing

Different objectives also lead to different management structures. An area of moorland aimed to provide factory workers with a pleasant day's walking at the weekend needs a very different type of management than an island rainforest containing the last remaining site of a rare species, or an area of savannah under intense poaching pressure, or a mountain slope that needs to keep its forest cover to help prevent snow from engulfing the village below.

There have been various evolving attempts to define a "protected area". We will dig down into some of the complexities that bedevil these issues. And international definitions are only a guide. Governments tends to interpret them as they see fit, and practical day-to-day management may not follow government rules anyway, so any list of protected areas will contain a myriad of different approaches within a broad spectrum of conservation effort.

The phrase "protected area" is itself problematic; people outside the conservation arena hardly recognise it at all. It can provoke an almost visceral dislike from many people in the extractive, forestry or agricultural industries and suggests a far from inclusive form of management for some of those actively protecting human rights. The concepts it embraces – national parks, nature reserves, wilderness areas, game reserves – are much more recognisable but also less precise (Dudley, 2008). A "national park" can vary between countries, from being an area strictly set aside for wildlife, landscapes and recreation as in North America and Africa, to a living cultural landscape with many settled human communities as is the case in much of Europe. Some places labelled "national parks" do not fit within our understanding of a protected area at all, being urban parks or similar. A "wilderness area" describes an approach to management that is passionately defended by a proportion of the conservation community, but the term "wilderness" provokes strong antipathy amongst other sectors, particularly indigenous people who point out that they have lived in and managed most such apparently "natural" areas for millennia (Callicot, 2000), and groups like farmers and foresters for whom wilderness holds negative connotations of mismanagement and abandonment. And the term "*private* protected areas", which has long been used to distinguish private conservation efforts from those of governments, was recently revised to "*privately* protected area" to better define an approach that uses private means for conservation, but can provide public goods, such as the increasing number of NGO reserves around the world which millions of people support and visit (Stolton et al., 2014).

To add to the complications, in the last few years a number of new concepts have emerged that fall outside the usual definitions of a protected area but nevertheless are places that play an important role, or potentially important role, in conserving natural ecosystems. Designations with the clumsy titles of "Other effective area-based conservation measures" or "connectivity conservation" and similar are increasingly recognised as conservation tools.

Terminology is still developing and a more general term of "area-based conservation" is emerging to describe the whole basket of approaches linked to management decisions that relate to particular sites. As mentioned in Chapter 1, this will be the term we commonly use here although new names may well emerge again in the near future; area-based conservation is after all another generally off-putting term that is unlikely to grab the public imagination. It does mark an important step forward in our collective understanding of land and water management. It embodies a new approach to global terrestrial and marine ecosystems that has implications at virtually every level of society and every industrial sector. We are only just starting to come to terms with what it will mean in practice. It is likely to become increasingly important as efforts to halt conversion of natural ecosystems intensify along the development frontier.

Even more significant than the evolving understanding about the management of area-based conservation is a parallel evolution in "who makes the decisions" about management. The modern protected area movement emerged on government-owned or government-requisitioned land; once a decision had been made to designate a protected area the practical implications were a matter for the state. But as soon as national parks started to be designated in Europe, where much of the land was privately owned, decision-making processes became more complicated. Over the last twenty years, two important developments have taken place. Within traditional state-owned protected areas there has been an increasing call for greater involvement by other rightsholders and stakeholders, including sometimes people who once had traditional ownership of the land, or other interested actors, or people still living within or near the area. Various forms of collaboration have emerged as a result, ranging from the minimal (some consultation) to experiments of joint management. In parallel, other groups or individuals who own or have management rights over land and water have decided to bring these areas into the conservation estate, either formally or informally. These range from indigenous peoples, through local communities to religious groups, companies, ecotourism ventures, charitable trusts and individual people. As with state-run protected areas, these community and privately protected areas started as a trickle and in some cases have now become a flood. Different governance structures are increasingly being recognised, both in new protected areas and, a far more complicated process, being retrofitted onto existing protected areas in response to public pressure or management failures. These fundamental changes about who is in charge, along with a deeper-seated understanding about the importance of governance, are creating a revolution not only in conservation but also in our understanding about how the world might be better managed in the future.

What does the future hold for protected areas?

So, the large majority of the world's protected areas have been established during our lifetime; indeed, established during the lifetime of many people way younger than us. But will they still be protected in fifty or a hundred years' time? This is a period of extraordinary change: on the one hand a huge upsurge of interest in the environment and concern that we have a tiny window of opportunity left to maintain a secure planet for future generations. But there has also been an emergence of right wing, authoritarian governments, fake news and an ideological attack on anyone labelled "green" or similar. The industrial complex straddles this debate; on the one hand some companies provide leadership and vision for a more sustainable approach in places where governments have apparently dropped the ball, while on the other hand powerful corporations are fighting an extraordinary rear-guard action in defence of short-term strategies that will further undermine the planet's ecosystems.

This is not to say that everyone who opposes a protected area is driven by nefarious motives. There are legitimate frustrations and injustices with conservation as there are in all other areas of human endeavour. Farmers whose families spent generations managing land may have very understandable reasons for not wishing to see it return to "nature". As we write, our village in Wales is surrounded by posters attacking a rewilding project (interestingly they say "yes to conservation, no to rewilding"). Protected area strategies need to consider where people are coming from instead of simply where conservationists think they should go.

References

Abu-Izzeddin, F. (2013) *Memoirs of a Cedar: A History of Deforestation, a Future of Conservation.* Shouf Biosphere Reserve, Lebanon, pp 105–107

Anon. (1972) 'Blueprint for survival'. *The Ecologist,* vol. 2, no. 1

Bagader, A.A., Al-Chirazi El-Sabbagh, A.T., As-Sayyid Al-Glayand, M. and Izzi-Deen Samarrai, M.Y. (1994) *Environmental Protection in Islam,* 2nd edition. IUCN Environmental Policy and Law Paper no. 20. IUCN, Gland, Switzerland

Balmford, A. (2012) *Wild Hope: In the Front Lines of Conservation Success.* The University of Chicago Press, Chicago, chapter 2, pp 19–40

Bertzky, B., Corrigan, C., Kemsey, J., Kenney, S., Ravilious, C., et al. (2012) *Protected Planet Report 2012: Tracking Progress Towards Global Targets for Protected Areas.* IUCN, Gland, Switzerland and UNEP-WCMC, Cambridge

Callicott, J.B. (2000) 'Contemporary criticisms of the received wilderness idea'. In: Cole, D.N., McCool, S.F., Freimund, W.A. and O'Loughlin, J. (eds.) *Wilderness Science in a Time of Change Conference – Volume 1: Changing Perspectives and Future Directions; 1999 May 23–27 Missoula, MT.* Proceedings RMRS-P-15-VOL-1. U.S. Department of Agriculture, Forest Service, Rocky Mountain Research Station Ogden, UT

Carruthers, J. (2012) 'National parks, civilisation and globalisation'. In: Gissibl, B., Höhler, S. and Kupper, P. (eds.) *Civilising Nature: National Parks in Global Historical Perspective.* Berghahn, New York and Oxford, pp 256–265

Choudhury, A. (2004) *Kaziranga: Wildlife in Assam*. Rupa and Co, New Delhi, pp 1–2

Dudley, N. (ed.) (2008) *Guidelines for Applying Protected Area Management Categories*. IUCN, Gland, Switzerland

Dudley, N., Buyck, C., Furuta, N., Pedrot, C., Bernard, F., et al. (2015) *Protected Areas as Tools for Disaster Risk Reduction: A Handbook for Practitioners*. IUCN and the Ministry of Environment, Japan

Duffy, R. (2010) *Nature Crime: How We're Getting Conservation Wrong*. Yale University Press, New Haven and London

Fitter, R. and Scott, P. (1978) *The Penitent Butchers: 75 Years of Wildlife Conservation*. Collins and the Fauna Preservation Society, London

Govan, H., Aalbersberg, W., Tawake, A. and Parks, J. (2008) *Locally-Managed Marine Areas: A Guide for Practitioners*. The Locally-Managed Marine Area Network, Suva, Fiji

Hey, D. (2011) 'Kinder Scout and the legend of the mass trespass'. *Agricultural History Review*, vol. 59, pp 199–216

Howard, P., Davenport, T.R.B., Kigenyi, F.W., Viskanic, P., Baltzer, M.C., et al. (2000) 'Protected area planning in the tropics: Uganda's national system of forest reserves'. *Conservation Biology*, vol. 14, no. 3, pp 858–875

Jones, K. (2012) 'Unpacking Yellowstone: The American national park in global perspective'. In: Gissibl, B., Höhler, S. and Kupper, P. (eds.) *Civilising Nature: National Parks in Global Historical Perspective*. Berghahn, New York and Oxford, p 35

Kümpel, N. (2014) *Global Public Opinion Survey on Space for Nature*. Presentation at the World Parks Congress, Sydney, Australia

Kunwar, K.J. (2009) *Four Years for the Rhino: An Experience of Anti-Poaching Operations*. Save the Rhino Foundation, Kathmandu

Mathur, V., Verma, A., Dudley, N., Stolton, S., Hockings, M. and James, R. (2007) 'Kaziranga national park and world heritage site, India: Taking the long view'. In: *World Heritage Reports no. 21 – World Heritage Forests*. UNESCO, Paris

McEwan, A. and McEwan, M. (1982) *National Parks: Conservation or Cosmetics?* George, Allen and Unwin, London

McShane, T.O. and McShane-Caluzi, E. (1997) 'Swiss forest use and biodiversity conservation'. In: Freese, C.H. (ed.) *Harvesting Wild Species: Implications for Biodiversity Conservation*. John Hopkins University Press, Baltimore and London

Meadows, D.H., Meadows, D.L., Randers, J. and Behrens III, W.W. (1972) *The Limits to Growth*. Universe Books and The Club of Rome, New York

Muir, J. (1901) *Our National Parks*. Houghton Mifflin Company, New York

Reid, J. (2013) 'Managing Australia's world heritage in the greater blue mountains'. In: Figgis, P., Leverington, A., Mackay, R., Maclean, A. and Valentine, P. (eds.) *Keeping the Outstanding Exceptional: The Future of World Heritage in Australia*. IUCN National Committee of Australia, Sydney

Royal Dutch Shell. (2013) Sustainability Report 2013

Saikia, A. (2009) 'The Kaziranga national park: Dynamics of social and political history'. *Conservation and Society*, vol. 7, pp 113–129

Schama, S. (1995) *Landscape and Memory*. Fontana Press, London, pp 37–53

Sheail, J. (2010) *Nature's Spectacle: The World's First National Parks and Protected Areas*. Earthscan, London, pp 23–24

Small, R. (ed.) (2016) *The Last of the Buffalo: Return to the Wild*. Summerthought Publishing, Banff, Canada

Stolton, S. and Dudley, N. (eds.) (2010) *Arguments for Protected Areas: Multiple Benefits for Conservation and Use*. Earthscan, London

Stolton, S., Redford, K.H. and Dudley, N. (2014) *The Futures of Privately Protected Areas*. IUCN, Gland, Switzerland

Verschuuren, B., Wild, R., McNeely, J. and Oviedo, G. (eds.) (2010) *Sacred Natural Sites: Conserving Nature and Culture*. Earthscan, London

Watson, J.E.M., Dudley, N., Hockings, M. and Segan, D. (2014) 'The performance and potential of protected areas'. *Nature*, vol. 515, pp 67–73

PART II

What has been achieved so far

4

AGREEING WHAT WE MEAN BY AREA-BASED CONSERVATION

After the early experiments with setting aside land for wildlife and scenery, the global community has developed a working philosophy of ecosystem conservation: what it means in terms of management and who makes the decisions. The term "area-based conservation" has emerged to embrace a wide variety of different management models on land, in freshwater and covering the world's coastal and oceanic areas. Alongside what we might call traditional "protected areas" like national parks and nature reserves, which over the last half century have evolved agreed definitions, management approaches and principles of use, there are a wide and widening variety of different initiatives and conservation models in various stages of development. The clumsily titled "other effective area-based conservation measures" is one of these. Some approaches are so new that we still don't really know how they will operate out in practice; others are still in the process of being worked out. Furthermore, the question of who takes decisions about location and management of protected areas is also changing and different governance models are being recognised and codified. Our concepts of what constitutes area-based conservation are currently changing more quickly than at any time in the last century.

Introduction

The easiest way to get into Serengeti National Park is to fly: a bumpy ride from Arusha due to the hot air rising from the plains, followed by an exhilarating landing as the pilot buzzes the runway to drive away any animals that might otherwise panic and run in front of the plane. Flying also helps to put the park into perspective against the wider landscape of Tanzania; Mount Kilimanjaro rising majestically up from the plain and at the right time of year a million wildebeest and zebra can be seen migrating, trotting steadily across the savannah. There are mountain ridges and depending on the way the flight goes, a chance to pass over the vast Ngorogoro Crater. Outside the park there are small homesteads, with circular stockades of thorn bushes to protect against predators, sadly often surrounded by eroded ravines and other signs of degradation. There are also areas of cropland and some large ranches. But what is increasingly most obvious is the boundary of the park itself: a stark line between "natural" ecosystem and "farmed" landscape. Despite many fine words about ecological connectivity and buffer zones around protected areas, much of the world's protected area system exists as islands in a sea of development. This is currently how many people view protected areas, itself a situation that we need to address.

It is not just that the world set up a system of protected areas in the last century, but it also had to decide exactly what that meant. This is not a small or insignificant task. For virtually the whole of human history, over much of the planet, the emphasis has overwhelmingly been on subduing nature rather than letting it run free: taming rivers, felling forests and ploughing up grassland. This approach is far from ancient history; many of the famers and foresters around where we live regard unmanaged nature with deep suspicion as wasted land and the term "wilderness" is regarded as derogatory by many land custodians. Foresters still talk about "over-aged" trees. Such attitudes are being played out on a larger scale at the moment in debates about the future of the Amazon, the Congo Basin and the Northern Great Plains of the United States.

Five years ago, if we had been writing a book about areas designated for conservation, the focus would almost certainly have been wholly on "protected areas" defined by international agreement and listed on the World Database on Protected Areas. Things have got a lot more complicated since, but we will start this brief survey by looking at protected areas, which remain by far the commonest conservation tool in terms of setting aside areas of land and water.

Defining protected areas

The concept of a "protected area" has developed gradually since the emergence of the modern conservation movement. It aims to provide a common language for a global movement made up of a mix of individual actions and country, species or biome wide initiatives. It embraces many terms like "national park", "nature reserve" and "wilderness area". Without wishing to get bogged down in a long history, there were several attempts at setting out what it means to be a protected area during the latter part of the twentieth century (see Bishop et al., 2004). The world now rather confusingly has two protected area definitions running in parallel, one from the United Nations' Convention on Biological Diversity (CBD) and the other from the International Union for Conservation of Nature (IUCN), agreed in 2008 after several years of consultation (Dudley, 2008).

The **CBD** defines a protected area as: "A geographically defined area which is designated or regulated and managed to achieve specific conservation objectives".

IUCN has a slightly different definition: "A clearly defined geographical space, recognised, dedicated and managed, through legal or other effective means, to achieve the long-term conservation of nature with associated ecosystem services and cultural values".

This is not ideal and gives some idea about how intensely these issues are debated; there is tacit agreement between the two institutions that the definitions are equivalent from an operational viewpoint and the IUCN definition is the standard against which data on protected areas worldwide are collected and distributed (Bingham et al., 2019). They encapsulate a large amount of information in a few words. Guidance from IUCN has provided explanation for most of the words and phrases; and a set of principles to guide interpretation. Given the huge areas of land and water involved, the details of precisely what is expected of a protected area system becomes important and we produce IUCN's guidance in shortened form in the following (see Dudley, 2008).

Clearly defined geographical space: includes land, inland water, marine and coastal areas or a combination. "Space" has three dimensions. For instance, the airspace above a protected area may or may not be protected from low-flying aircraft, or there might be cases in marine protected areas when water is protected only to a certain depth or the seabed is protected but water above is not. Subsurface areas are sometimes not protected (e.g., are open for mining) even in otherwise strictly protected areas. "Clearly defined" implies a spatially defined area with agreed and demarcated borders. These borders can sometimes be based around mobile physical features (e.g., riverbanks or shorelines) or by management actions (e.g., agreed no-take zones).

Recognised: implies that all such sites should be recognised in some way (in particular through listing on the World Database on Protected Areas and the *UN List of Protected Areas*).

Dedicated: implies commitment to conservation in the long term, through for example international conventions and agreements; national, provincial and local law; customary law; covenants of NGOs; private trusts and company policies; or certification schemes.

Managed: assumes active steps have been taken to conserve the natural (and possibly other) values of the protected area. Note that "managed" can include a decision to leave the area untouched; many protected areas in northern Australia for instance have no active management but are also under no particular pressure.

Legal or other effective means: implies that protected areas must either be gazetted (that is, recognised under statutory civil law), recognised through an international convention or agreement such as UNESCO World Heritage, or managed through other effective but non-gazetted means, such as through recognised traditional rules under which community conserved areas operate, or the policies of established non-governmental organisations. Recognition of the importance of community agreements and traditional rules as legitimate ways of establishing protected areas has increased over the last few years. In Switzerland, for example, members of a canton inside a protected landscape, such as the Jurassic Park in the Jura Mountains, have the periodic right to vote about whether to continue protected area status. Similarly, although some privately protected areas fall under national legislation, e.g. in South Africa and Brazil, many do not and thus defining "effective means" focuses on processes to recognise that the area meets the protected area definition (Stolton et al., 2014).

. . . **to achieve**: implies some level of effectiveness. Management effectiveness is progressively being recorded on the World Database on Protected Areas and over time will become an important contributory criterion in identification and recognition of protected areas.

Long-term: protected areas should be managed in perpetuity and not as a short-term or temporary strategy. "Intent" is important here; some jurisdictions insist on periodic reviews of protection status, as in the Swiss example earlier, while others are based on funding mechanisms or agreements, such as conservation covenants, which are subject to a specific time period; but to be a protected area the *aim* should be conservation in perpetuity.

Conservation: refers to the *in-situ* maintenance of ecosystems and natural and semi-natural habitats and of viable populations of species in their natural surroundings and, in the case of domesticated or cultivated species, in the surroundings where they have developed their distinctive properties. Conservation often implies active management.

Nature: in this context nature *always* refers to biodiversity, at the genetic, species and ecosystem level, and often *also* refers to geodiversity, landform and broader natural values. The island of Rum, in Scotland, was originally established as a National Nature Reserve because of its unique geological features but is also the site of a globally important colony of Manx shearwaters (*Puffinus puffinus*) (Clutton-Brock and Ball, 1987). "Nature" also has more complex connotations for many indigenous and traditional communities, who include within the word aspects of the sacred and of human interaction with the wider world.

Associated ecosystem services: means those that are related to but do not interfere with nature conservation. These can include provisioning services such as contributions to food and water security; regulating services including regulation of floods and drought; supporting services such as nutrient cycling; and cultural services that can range over recreational, spiritual and religious benefits (Dudley et al., 2006).

Cultural values: includes any that do not interfere with the conservation outcome. They can sometimes contribute positively to conservation (e.g., traditional management upon which key species have become reliant); or be themselves under threat, such as traditional woodland management like coppicing, or maintenance of sacred natural sites.

This breakdown demonstrates the very wide range of issues that have been discussed, exhaustively, in the development of the global protected area estate. It should be noted that the definitions from the CBD and IUCN are "guidelines". The CBD has some political influence due to the fact that most countries of the world are signatories, while IUCN relies solely on the agreement of its government and non-governmental members. The details of what does and does not "count" as a protected area are determined by national policy and laws; some countries report areas as protected areas that do not meet, or only dubiously meet, the IUCN or

CBD definition. Nor will this be the end of the story; the "new" IUCN definition was agreed in 2008, and there are already voices calling for revision (e.g., Lee, 2015).

Consequently, there is much more variation in the global protected area system than a casual glance at the maps on the Protected Planet website might suggest. Our own experience however is that most governments do not want to be seen to be "breaking the rules" to any significant degree and the agreed global definitions carry weight beyond their lack of legal standing. Unfortunately, as the types of management and governance become more varied, the ability of governments to report on area-based conservation is declining, in part because sources of information are not always clear.

One long-standing question addressed during the 2008 revision of the IUCN definition was whether nature conservation was always the main aim of a protected area or if it could be superseded in some cases by values such as wilderness protection, landscape protection and recreation. This was a long and sometimes bitterly contested issue that was finally resolved, at least for now, by a show of hands at an IUCN workshop in Almeria, Spain (Stolton and Dudley, 2009). The consensus was eventually, and at least within the conservation community overwhelmingly, that nature conservation should be the priority. But the ramifications of this debate rumble on, with strong opposition from some of the Pacific island states for instance, and these discussions may in turn have helped spark some of the new approaches discussed later.

The IUCN definition is thus further clarified by the addition of a series of complementary principles, the most important in this context being:

> For IUCN, only those areas where the main objective is conserving nature can be considered protected areas; this can include many areas with other goals as well, at the same level, but in the case of conflict, nature conservation will be the priority.
>
> *(Dudley, 2008)*

This recognises that many protected areas will have other important management priorities – cultural, spiritual, tourist-related and so on – but that for a place to be a protected area recognised by IUCN and included in the *UN List of Protected Areas*, conservation needs to take priority. It is in marked contrast with some of the approaches to area-based conservation discussed later in the chapter.

Other principles refer to such things as the value of all types of protected area to conservation, management effectiveness, selecting the best management approach for a particular situation, and one critical statement relating to human rights: "The definition and categories of protected areas should not be used as an excuse for dispossessing people of their land" (Dudley, 2008). The extent to which this has been met in practice will be discussed in Chapter 11.

Another important and practical aspect of the IUCN protected area definition is the so-called "75 per cent rule": this states that up to 25 per cent of land or water in a protected area can be managed for other purposes so long as these are compatible with its primary objective. This is to accommodate existing villages and houses, tourist lodges, fishing areas for local communities and, conversely, also applies to small strictly protected reserves within protected landscapes and seascapes. The exempted area may sometime be movable.

Various types of protected areas

Both IUCN and the CBD recognise a range of management approaches as being applicable within protected areas, as long as the area meets the overall definition of a protected area. In other words, if the basic objectives are met, primarily nature conservation but also ecosystem services and cultural values, there is a fair amount of latitude about the ways in which this is achieved. Table 4.1 summarises six management categories that can apply to protected areas. Unlike the definition, these are recognised by both the CBD and IUCN, and through this system we also have a common language to discuss the concept of protected area management around the world. We were recently in Rwanda and hired a local ranger in Nyungwe National Park for a day's bird watching. The birds proved rather elusive and a thunderstorm curtailed the trip, but of relevance here is that in explaining the work of the park the ranger immediately referred to its "Category II" status and what it meant in terms of management, including the many ways that the park was working with local communities to improve livelihoods but maintain the forest. A common standard not only helps those responsible for protected areas understand where their particular site sits within the overall conservation matrix, but also helps to provide peer-pressure to ensure that different areas aspire to the same overall standards.

Most protected areas are assigned to a particular category by governments, often by calibrating national systems to the IUCN system for international reporting. Some very large protected areas, where different legally established zones use markedly different management approaches, can have more than one category: The Great Barrier Reef offshore of Queensland, Australia, is one such example.

A glance at the table shows that protected areas exist under a wide range of management approaches. Bwindi Impenetrable National Park is close to the border in Uganda, butting up to both Rwanda and Congo Brazzaville, and covering 320 km^2. It is a dense area of forest-covered volcanoes and home to a large proportion of the world's surviving mountain gorillas (*Gorilla beringei beringei*), with a full-time research station inside. It is a magnet for visitors, who are willing to pay $1,000 or more for a chance to trek to one of a handful of habituated gorilla family groups. Access is strictly controlled. Most visitors only stay long enough to see the gorillas and then head off to other national parks or back to the modern hotels of Kampala. Bwindi also plays an important role as a water tower in the country, but its role and future are far from uncontested. The rich volcanic soils, warm weather and abundant rainfall produce rich crops on steep-sided terraces; like Serengeti its borders are mainly hard edged as farmers squeeze up to the forest, surreptitiously clearing a few extra metres if they get the chance. The batwa (pygmy) people of the area complain they have lost traditional territory to the protected area. Rangers are

TABLE 4.1 IUCN and the CBD recognise several different management categories

Ia	Strictly protected areas set aside to protect biodiversity and also possibly geological/ geomorphological features, where human visitation, use and impacts are strictly controlled and limited to ensure protection of the conservation values. Such areas can serve as indispensable reference areas for scientific research and monitoring.
Ib	Usually large unmodified or slightly modified protected areas, retaining their natural character and influence, without permanent or significant human habitation, which are protected and managed so as to preserve their natural condition
II	Large natural or near natural areas set aside to protect large-scale ecological processes, along with the complement of species and ecosystems characteristic of the area, which also provide a foundation for environmentally and culturally compatible spiritual, scientific, educational, recreational and visitor opportunities
III	Set aside to protect a specific natural monument, which can be a landform, sea mount, submarine cavern, geological feature such as a cave or even a living feature such as an ancient grove. They are generally quite small protected areas and often have high visitor value.
IV	Aim to protect particular species or habitats and management reflects this priority. Many Category IV protected areas will need regular, active interventions to address the requirements of particular species or to maintain habitats, but this is not a requirement of the category.
V	An area where the interaction of people and nature over time has produced an area of distinct character with significant ecological, biological, cultural and scenic value, and where safeguarding the integrity of this interaction is vital to protecting and sustaining the area and its associated nature conservation and other values
VI	Conserve ecosystems and habitats, together with associated cultural values and traditional natural resource management systems. They are generally large, with most of the area in a natural condition, where a proportion is under sustainable natural resource management and where low-level non-industrial use of natural resources compatible with nature conservation is seen as one of the main aims of the area.

armed, patrol in large groups and have to deal with poachers, incursions and, for many years, rebels from the Rwandan genocide camped out in the forests. Several tourists were abducted and murdered in 1999, which had a devastating impact on income for a number of years; Bwindi is normally the cash cow that keeps the rest of the Uganda Wildlife Service in operation. Bwindi has been assigned management Category II and thus is a fairly strictly protected area and is recognised by UNESCO as a natural World Heritage site.

Compare Bwindi with the Lake District National Park, a protected landscape in the UK, consisting of many mountains, woodland, towns and villages and the numerous lakes that give the area its name. The national park covers over 2,300 km², almost an order of magnitude larger than Bwindi, but mainly in private hands, predominantly upland sheep farming, also some state and private forestry and a scatter of more strictly protected nature reserves. But despite the differences, like Bwindi, the main source of income today is tourism with over 15 million visitors a year, although land-use decisions are still dominated by farming. The Lake District is also under pressure but of very different kinds; heavy grazing in the uplands has created large areas of short grass sward with little absorptive capacity, leading to increased flooding in the valleys below. There has recently been an intense public debate about the use of off-road motorised vehicles in tracks within the park. Rangers deal predominantly with visitors. Several wild plant and animal species are declining due to long-term land use change and tourism pressure is having impacts, particularly through erosion of mountain paths. Tension between farming and tourism provides a background to management, with many upland farmers resentful of the number of visitors. The Lake District is recorded as IUCN Category V, a protected landscape. In 2015 it also received UNESCO World Heritage status, but significantly this was not because of its natural values, but rather its cultural links with the Romantic movement and poets such as William Wordsworth. Indeed, there was quite a backlash from conservationists about the designation, with George Monbiot describing the Lakes as a "shipwrecked monument to subsidised overgrazing and ecological destruction" (2017).

The differences are not simply those between a developed and a developing country, between rich and poor. Sarek National Park in Sweden is almost 2,000 km² in size, far north in the Arctic Circle and only accessible from the nearest town by boat. It contains almost nothing by way of settled communities, although the indigenous Sámi people live there while herding reindeer during the summer and there are a series of cabins for hikers on the two-week trail that runs through the park. Northern Scandinavia is one of the few places with the space and low population to emulate the kind of "wilderness experience" beloved of North American adventurers, and protected area agencies have reached a trade-off with the Sámi so that the national park system both protects the ecosystem and sets aside enough land to allow traditional reindeer herding to continue. When we went there, early in the season, we camped and walked for several days and didn't see another person, a rare opportunity anywhere, let alone in Europe.

Compare this in turn with Brijuni National Park in Croatia, where annually over 100,000 visitors travel around on a toy train to see a safari park full of exotic animals. After World War II, the main island became the summer residence of President Tito of Yugoslavia; over a hundred foreign heads of state visited the president on the island, many bringing gifts of animals which have outlived both the president and his country. Visiting Brijuni was initially a shock – how could this possibly be a protected area? – until it was explained that this rather sad relic is just one of fourteen islands, and 80 per cent of the park protects the rich marine ecosystem of the Adriatic.

All of these sites are protected areas recognised by IUCN, the CBD and the international community and listed on the World Database of Protected Areas. Bwindi, Brijuna and the Lake District all rely predominantly on tourism to support local communities; in Sarek there is a mixture of reindeer herding and low-level tourism.

Although all are national parks, they are dramatically different in management, the pressures they face, in aspirations and objectives and in who owns and makes decisions about the land. Both approaches – strict protection and the protected landscape approach – have their passionate supporters and also their detractors.

International designations of protected areas

As noted, Bwindi and the Lake District are both also UNESCO World Heritage sites, one of a number of international designations of protected areas that are applied to certain selected and approved protected areas and normally sit alongside national designations. Three are important and together help to define management over a significant area of the planet.

UNESCO World Heritage

The World Heritage Convention was established to help protect natural and cultural heritage. Today its main focus is the World Heritage List, an approved list of buildings, other cultural sites and natural areas with the objectives of representing examples of the world's most important heritage. There are now over a thousand sites on the list, of which most are cultural, predominantly buildings but also some landscapes of cultural value. The coffee-growing region of Colombia is an example of the latter. There are in addition just over 200 natural sites and around forty mixed sites (with both natural and cultural values). Because sites serve as examples, each new site listed is supposed to display Outstanding Universal Value (OUV), a hard-to-define attribute which basically means that nothing too similar has been included on the World Heritage list before. There is a lengthy application process, site visits by specialists and a vote by the World Heritage committee; it is by now unfortunately an increasingly politicised procedure. The four natural criteria extend beyond conventional nature conservation concerns and include natural beauty, geological importance and biological processes alongside biodiversity. Most countries of any size have at least one World Heritage site while a few states apply for new sites every year and are building a substantial list. The natural and mixed sites already represent examples of some of the most iconic ecosystems on the planet: Serengeti, the Grand Canyon, Iguazu, the Great Barrier Reef, Central Amazon, Komodo National Park and so on. Their high importance means that they are supposed to be amongst the best managed; unfortunately, this is not invariably the case. Where the site's integrity is at serious risk, the World Heritage Committee can agree to list the area as World Heritage in Danger; this has no legal sanctions but increases international pressure on a country to respond and can help unlock additional funding assistance.

Volcanic remnants in South Korea

Jeju Island, in South Korea, is a typical case of natural World Heritage. A hugely popular destination for Korean tourists, with the coastal area filled with hotels and golf courses, the centre of the island is dominated by the volcanic Mount Halla, which is within a larger national park. The World Heritage nomination includes the national park and also Seongsan Ilchubong tuff cone, a volcanic plug rising out of the ocean like a fortress and connected to the mainland by a causeway; the Geomunoreum lava tubes, regarded as the finest in the world; and cliffs of basalt columns like those of the Giant's Causeway in Ireland. The tuff cone and those lava tubes open to the public are packed with visitors for most of the year – no wilderness experience here. Most of Mount Halla is off limits, although there are heavily used hiking paths scattered throughout the park and a plush visitors' centre. The government has invested a lot of money in both protecting and promoting the World Heritage site, with all but a trivial number of visitors being domestic. As a model for World Heritage it illustrates both the strengths and paradoxes of the list: the central protected area is indeed protected from the development that has affected much of the rest of the island, and World Heritage listing has highlighted the unique values and attracted many tourists. The tourist boom means that parts of the site can only be experienced in crowds not far short of what might be expected in Notre Dame or the Taj Mahal. We have visited the central protected area several times both professionally and for pleasure, and there is still the opportunity, just about, to get away from the noise and bustle of walking tours and enjoy some quiet bird watching and contemplation. Maybe in a crowded planet, that is part of the trade-off for having unique values? **SS/ND**

UNESCO's Man and the Biosphere

Surprisingly, UNESCO is also home to one of the other three most important "international" protected area designations, despite there being several other United Nations agencies with a far more obvious role in environment. UNESCO's Man and the Biosphere programme has built a parallel list of MAB biosphere reserves, with a very different conservation philosophy to World Heritage. Biosphere reserves are based around three management approaches: (i) core zones where biodiversity protection is given priority (there may be several of these in any biosphere reserve); (ii) buffer zones, where management is aimed to support conservation in the core zone(s) whilst allowing activities such as farming and forestry; and (iii) a transition zone, a larger area where people live and work and conservation activities are focused on education, training and so on. The biosphere reserve is one of the most "modern" approaches to area-based conservation in that it consciously integrates conservation and human activities (Hadley, 2002).

There are over 700 biosphere reserves around the world. A number of countries have used the designation imaginatively, working with local communities to develop sustainable management approaches in the buffer zones and carrying out high standards of conservation and restoration in the core (Stoll-Kleeman et al.,

2010). Some of Lebanon's almost vanished cedars are now being preserved and restored in Al Shouf biosphere reserve for instance (Khalil and Hani, 2014). However, many countries have struggled with the application, and all too many biosphere reserves are little more than the core reserves (which are not always even that well managed) with buffer and transition zones that have little or no relevance in practice.

Cat Ba Biosphere Reserve, Viet Nam

Cat Ba Island in Viet Nam is at the edge of the famous Ha Long Bay World Heritage site, with its pillars of rock jutting up from the sea. The reserve covers 90 per cent of the island, surrounded by a marine area with 366 smaller islands. Its core zone includes a national park (IUCN Category II) and marine park (IUCN Category V), with subtropical rainforest on rugged limestone mountains, wetlands, mangroves, seagrass beds and coral reefs. There are endemic species including the Cat Ba leopard gecko (*Goniurosaurus catbaensis*), the Cat Ba langur (*Trachypithecus poliocephalus poliocephalus*), one of the most threatened primates in the world, and the Ha Long palm (*Livistona halogensi*). The langur's population was estimated at 2,400–2,700 individuals in the 1960s but fell to around 50 by 2000 due to hunting for the medicine trade (Stenke and Chu, 2004). Numbers have stabilised since, due to work by the Cat Ba Langur Conservation Project through capacity-building and education activities and now stands at 64. But pressures remain and an entire group of langurs were poached in 2015 (Tatarski, 2019).

Cat Ba Island is also a booming tourist destination, now hosting over 2.5 million visitors a year, mainly domestic. The social structure is being transformed in the process although many people still live primarily by fishing or agriculture. Seafood farming enterprises, including coastal clam, seafood ponds and fish-cage farming boomed after aquaculture was banned in the Ha Long Bay World Heritage area. The construction of many industrial-sized seafood ponds on the main island, and around 12,000 floating live-aboard fish farms in the marine areas (including the core zone of the biosphere reserve) generate income but add to development pressure (Le Thanh Tuyen, 2011). In 2019, rumours of a US$3 billion new tourism development on a neighbouring island emerged, with plans to link to Cat Ba by cable car. Awareness of the biosphere reserve is high due to a strong publicity drive and promotion of an official Cat Ba Biosphere Reserve label for products and services that meet sustainability criteria. There have been attempts to restore mangroves, and these are reported to be in better condition than when I visited a decade ago. There is the beginning of ecotourism, but the majority of visitors come for the beaches rather than the wider environment. While the terrestrial core zone remains untouched by most tourists due to its sheer inaccessibility, these barriers do not deter poachers, and controls in the buffer zones seem fairly minimal. **ND**

Ramsar sites

The last of the international designations comes from the International Convention on Wetlands, usually known as the Ramsar Convention after the town in Iran where it was first signed in 1971. Unusually, the convention is an agreement between nations outside the auspices of the United Nations. Amongst its activities is the identification and designation of Ramsar sites, which are freshwater, estuarine or coastal wetlands judged to be of international importance from a conservation perspective and where governments make commitments to the concept of "wise use". There are over 2,300 such sites in total covering over 2 million km², ranging from tiny pools to huge areas of wetland. The Du-ung wetland in the Republic of Korea is less than seven hectares in size and is an example of a rare coastal freshwater lagoon fed by underground springs. Situated between coastal sand dunes and the mountains it appears unimpressive until its hydrological significance is explained. At another extreme, the Mamirauá site in the Brazilian Amazon covers over a million hectares of seasonally flooded varzea forest, with several lakes intermittently connected by drainage canals and high levels of endemism. Like World Heritage, Ramsar sites at risk can also be identified on the "Montreux list".

The relationship of all these sites to protected areas as described earlier is complex. Nowadays a natural World Heritage site is only accepted onto the list if it is already a protected area, but this was not always the case and some natural World Heritage sites are still not recognised as protected areas by the government concerned. Cultural landscapes will not all be protected; the coffee growing region of Colombia is undergoing serious deforestation on steep slopes as farmers seek to cash in on a growing demand for the region's coffee (which has probably been boosted by its World Heritage status). Biosphere reserves by their nature include areas that both are and are not strictly protected areas. And whether or not Ramsar designation is automatically equivalent to protected area status is a matter of debate even within Ramsar members; again, governments differ in their interpretation and the overall aim of Ramsar sites is for "wise use" rather than conservation as such.

Who is in charge?

Just as important – perhaps more important – than what type of management takes place inside a protected area is who ultimately makes the decisions; whether it is a government, a community, individuals or indigenous peoples' group for instance, or several different stakeholders working collaboratively. As with definitions and management approaches, understanding of governance in protected areas has developed over time.

When the modern protected area movement began all national parks were on land owned or requisitioned by the government and this remains true for the majority of the area of land and water under protection today. But things are changing; many more players are becoming actively involved in conservation, and at the same time other voices are demanding a hearing in state-managed protected areas, leading to a range of collaborative management models being developed.

For ease of understanding in an increasingly complex situation, IUCN has defined a typology of governance types – a description of who holds authority and responsibility for the protected area – to run alongside the management categories discussed earlier. These are summarised in Table 4.2 (Borrini-Feyerabend et al., 2012).

Each contains several sub-categories. Governance by governments can act at a number of levels: by the national government, or sub-national units such as states, provinces or counties, right down to quite a local level. The national government will generally retain overall decision-making power about what constitutes a

TABLE 4.2 Governance types of protected area

Type	Name	Description
A	Governance by government	A government body (such as a Ministry or Park Agency reporting directly to the government) manages the protected area and determines its management aims and objectives: ✓ Federal or national ministry/agency in charge ✓ Sub-national ministry/agency in charge ✓ Government-delegated management (e.g. to NGO)
B	Shared governance	Complex institutional mechanisms and processes are employed to share management authority and responsibility among a plurality of (formally and informally) entitled governmental and non-governmental actors: ✓ Collaborative management (various degrees of influence) ✓ Joint management (pluralist management board) ✓ Transboundary management (various levels over frontiers)
C	Private governance	Protected areas under individual, cooperative, NGO or corporate control and/or ownership set up and managed under not-for-profit or for-profit schemes: ✓ By individual owner ✓ By non-profit organisations (NGOs, universities, cooperatives) ✓ By for-profit organisations (individuals or corporate)
D	Governance by indigenous peoples and local communities	Protected areas under community or indigenous control: ✓ Indigenous peoples' conserved areas and territories ✓ Community conserved areas – declared and run by local communities

protected area and what rights other levels of government have in terms of setting up protected areas, but apart from that many decisions are increasingly devolved. Depending on the strength of governance inside a country, local and national authorities collaborate in planning and management of their respective protected areas to a varying degree.

Governments also sometimes retain overall decision-making power but hand over day-to-day running to non-governmental organisations. The Seychelles Island Foundation manages Aldabra Atoll and Vallée de Mai World Heritage sites on behalf of the government of the Seychelles for example. African Parks, a non-governmental organisation, has taken over running several protected areas in nine African countries, under long-term agreements with government. This has included some parks in particularly challenging situations, in areas of civil conflict, such as Chinko in the Central African Republic and Odzala-Kokoua in Congo and Garamba in the Democratic Republic of Congo.

Shared governance is a complex term that can embrace everything from occasional consultation with other stakeholders and rightsholders to complete joint decision-making. It increasingly also involves formal or informal agreements between adjoining protected areas across national borders (transboundary protected areas).

Although it is impossible to draw sharp distinctions, collaborative management can be divided into a series of broad stages. In the context of a protected area, the simplest approach is active consultation, involving the manager or management group in gathering input from a range of stakeholders and rightsholders, and using the results of the consultation to inform decisions. This can be given more weight by actively seeking consensus from those consulted, along with trying to ensure that no major interests are left out of the consultation process. The next stage is to introduce a process of negotiation and the development of specific agreements. Authority can also be shared in a more formal way through for example inviting or electing representative groups of rightsholders and stakeholders onto a management body. Finally, full authority is in some cases transferred to other rightsholders and stakeholders, so that the original governing body becomes just one of a series of voices in determining management.

What are stakeholders and rightsholders?

Stakeholders possess direct or indirect interests and concerns about a certain asset, state of things, actions or non-actions, but do not necessarily enjoy a morally, legally or socially recognised entitlement to them (Borrini-Feyerabend and Hill, 2015). *Stakeholders* can include individuals, families, community groups, government groups, businesses, religious groups, development agencies, NGOs; in fact, anyone with a legitimate interest in an issue. They do not need, necessarily, to be resident or even nearby. Millions of people around the world are concerned about the fate of species like the lion and tiger, even if they live in countries where these animals do not exist and have never seen one in their lives; they are stakeholders in the survival of these animals. It therefore follows that not all stakeholders are equal or have equally strong rights. In terms of ecosystems and species, it is generally assumed that those who live closer to or are more dependent on a particular resource have stronger rights than those who live further away or have only vague emotional connections to the same resource, although this doesn't mean that local desires invariably outcompete global needs. More recently the term rightsholders has been recognised to distinguish those groups or people who have relatively stronger claims over a particular issue or resource. A *rightsholder* is a person holding a right, i.e., whose interests are sufficiently important to raise a legal/moral duty (or set of duties) on one or more other persons (Raz, 1988). In the context of protected and conserved areas, rightsholders are those people and groups socially endowed with legal or customary rights with respect to land, water and natural resources.

(Borrini-Feyerabend and Hill, 2015)

These distinctions are, as stated earlier, far from precise: co-management is likely to be more in the nature of a continual problem-solving process rather than a fixed procedure (Carlsson and Berkes, 2005). In most cases the interaction will be between a state body and local stakeholders, but other permutations exist. Non-profit or for-profit privately protected areas may also wish to bring a wider range of stakeholders into the management process, such as local resource users, hoteliers, guides, research institutions and others with an economic stake in the future of the area or a long-standing interest in the way that the area is managed. More and more examples of co-management in area-based conservation are emerging, drawing on a wide range of models (Borrini-Feyerabend et al., 2004). However, contemporary analysis of prevalence, rates of success and failure, lessons learned, and of the management models used is strangely absent and there is clearly a deal more work needed in understanding how co-management models are being applied around the world.

"Privately protected areas" include those run by non-profit trusts; for-profit companies running ecotourism ventures on their own land; and a confusing jumble of individual owners, companies that set aside parts of their land, religious orders, universities and research groups (Stolton et al., 2014). In the UK, for example, the Royal Society for the Protection of Birds has over a million subscribing members and more than 170 reserves around the country. Leighton Moss is a case in point. Managed by the RSPB since 1964, although only covering 28 km² it contains the largest reed beds in northwest England and harbours populations of important wetland birds; it gained Ramsar listing in 1985. Like many small reserves, the area is managed for wildlife rather than being left alone (Wilson, 2005); small sites often do not have the space to develop natural dynamics. Reed beds are cut, managed to prevent them drying out and to stop saline intrusion from the nearby coast, and 200 hectares of extra reed bed have been created or restored. Visitors follow well-marked trails, and there are hides, information about the kinds of birds likely to be seen and when we were last there a display (a "murmuration") of tens of thousands of over-wintering starlings doing a spectacular aerial dance before roosting.

In the United States, The Nature Conservancy, with chapters in each state and a large global programme, estimates that it has helped protect more than 48 million hectares, including large areas of land that it owns. Privately protected area networks exist in many Latin American countries, often coordinating through their own networks, in sub-Saharan Africa, Australasia and Western Europe; they remain much less common in Asia, North Africa, the Middle East and in Eastern Europe, although they are gradually developing in all these places.

New players are emerging, or at least being recognised. All the world's major religions today recognise the important role of humans in acting as stewards of nature (Palmer and Finlay, 2003), and an increasing number of religious groups are actively managing their land as protected areas, or have buildings or sites *within* protected areas and in consequence are working with protected area authorities (Mallarach and Papayannis, 2006). In Europe, monks in several Christian monasteries within national parks are actively managing their lands for conservation including in the Montserrat National Park in Catalonia, Spain, Rila National Park in Bulgaria and lands around the monastery at Mount Athos in Greece. In Japan and South Korea, many temples exist within protected areas, so that a proportion of visitors come for mainly religious reasons. In Ethiopia, most of the remaining fragments of natural forest still surviving in large areas are in Christian churchyards (Aerts et al., 2006).

Commercial companies are also setting aside land, either because senior management is interested in conservation or as a result of voluntary certification schemes that stipulate protection for high conservation value areas (Stolton et al., 1999). For example, a Ramsar site in Chile, Santuario de la Naturaleza Laguna Conchalí, is owned and managed by Minera Los Pelambres, a copper mining company; their original environmental permit in 1997 stipulated that the wetland area should be protected. The reserve is a brackish coastal lagoon and a key area for birds migrating along the central Chilean coast. The Swedish-Finnish company Stora Enso owns and manages some of the last remaining fragments of natural forest recognized within the UNESCO World Heritage site in the threatened Atlantic Forests of Brazil. There are many other examples (Stolton and Dudley, 2008). Universities and similar institutions have also long owned and managed strict nature reserves for research, without necessarily viewing these as part of national protected area networks. The Andrews Experimental Forest in Oregon is a well-known example (Luyoma, 1999).

Private conservation makes up a tiny proportion of the world's total protected areas but often focuses on critical places and on sites connecting other protected areas. Just a couple of the many examples that we know ourselves include a private reserve protecting a leopard migration route in Armenia and Handa Island off the coast of northern Scotland, run by the Scottish Wildlife Trust and maintaining a unique seabird breeding colony. Direct land purchase can be a far quicker way to respond to a sudden increase in conversion of a sensitive habitat or the discovery of a population of an endangered species than waiting for slow-moving government decisions (Stolton et al., 2014). The importance of private conservation is likely to increase as state authorities reach the limit of areas that they are prepared to set aside.

South Africa is one of the leaders of such initiatives. Formal declaration of protected areas is through national legislation on state or privately-owned land. Currently 35 per cent of terrestrial protected areas are privately owned and 5 per cent communally owned through mechanisms such as Contract National Parks and Nature Reserves which protect sites with the highest biodiversity value and ecological infrastructure (Mitchell et al., 2018a).

Privately protected areas are likely to be particularly important for preserving fragments of habitat in heavily converted areas in tropical areas, which are often the last redoubts of particular plant species. In Chile, for example, public protected areas tend to be concentrated in the less populated areas of the extreme north and south of the country. The biodiversity hotspot of the Mediterranean Matorral contains the majority of Chile's population and includes the main agricultural region; not surprisingly it is the least protected part of the county. But privately protected areas are helping fill this gap and represent the primary form of protection in the region.

Finally, but importantly, there is a growing area of land and water managed directly by local communities and various communities or organisations comprised of indigenous peoples. "Territories and areas conserved by indigenous peoples and local communities", popularly called ICCAs, are a heterogeneous collection of places that are consciously managed with conservation in mind, often using management techniques dating back for hundreds or thousands of years. An ICCA is described as "natural and modified ecosystems including significant biodiversity, ecological services and cultural values voluntarily conserved by indigenous and local communities through customary laws or other effective means" (Borrini-Feyerabend, 2007). They may include places with minimal intervention but some sustainable off-take, such as areas of tropical forest inhabited by indigenous people who have an intimate knowledge of the flora and fauna, or quite intensively managed areas of land such as sacred natural sites or planted fruit gardens in tropical areas that nonetheless maintain high biodiversity value.

ICCAs contain examples of both the oldest and newest forms of protection. Some management systems have existed for hundreds or thousands of years and are now being "recognised" by the outside world. Others have developed more recently; many indigenous people have been forced to move in recently decades, due to persecution, political unrest or environmental change. ICCAs are also increasingly being incorporated within national conservation plans and protected area systems.

The following three characteristics identify an ICCA. First, there is a close and deep connection between a territory or area and an indigenous people or local community. This relationship is generally embedded in history, social and cultural identity, spirituality and/or people's reliance on the territory for their material and non-material wellbeing. Second, the custodian people or community makes and enforces decisions and rules (e.g., regarding access and use) about the territory, area or species' habitat through a functioning governance institution. And finally, the governance decisions and management efforts of the concerned people or community contribute to the conservation of nature (ecosystems, habitats, species, natural resources) as well as to community wellbeing. In the last few years, ICCAs have been provided with increasing levels of recognition and support at the international level, in large part through the ICCA Consortium, a grouping of indigenous people and their supporters committed to both human rights and environmental protection.

One of the most extensive networks of ICCAs is in Australia. Indigenous Protected Areas (IPAs) are areas of land and sea managed by Australian aboriginal groups as protected areas through voluntary agreements with the government; these often cover huge areas. There are currently seventy-five IPAs over 67 million hectares, which make up about 44 per cent of the country's National Reserve System. More IPAs are being developed, and soon about half of the National System will be IPAs. Vital for conservation, IPAs also help protect cultural values (Clarke and O'Leary, 2018).

Area-based conservation beyond protected areas

In 2010, the whole concept of what is and is not a protected area was thrown up into the air again. At a Conference of Parties of the CBD, held in Sendai, Japan, a ten-year set of biodiversity targets were agreed. Named after the Japanese prefecture in which the meeting took place, the "Aichi biodiversity targets" set much of the international agenda on biodiversity conservation for a decade. In the small hours of the morning, exhausted delegates arguing about the amount of the planet's surface that should be in protected areas invented a new phrase within Aichi Target 11 and started a decade of debate about its implications: "By 2020, at least 17 per cent of terrestrial and inland water areas and 10 per cent of coastal and marine areas . . . are conserved through . . . systems of protected areas *and other effective area-based conservation measures*" (our emphasis).

IUCN and the CBD Secretariat initially argued (Lopoukhine and de Souza Ferreira Dias, 2012) that this was virtually the same as a protected area and should be treated as such (Woodley et al., 2012), but this was rejected and IUCN was requested to assist in defining an "other effective area-based conservation measure" or OECM. A task force was set up and after long debate and consultation produced draft guidance for the CBD. After further editing, signatory countries agreed a definition in November 2018 at the 14th Conference of Parties in Sharm el Sheik, Egypt (CBD, 2018):

> A geographically defined area other than a Protected Area, which is governed and managed in ways that achieve positive and sustained long-term outcomes for the in-situ conservation of biodiversity, with associated

ecosystem functions and services and where applicable, cultural, spiritual, socio-economic, and other locally relevant values.

This covers three main situations:

1   **"Primary conservation"** areas that meet the IUCN definition of a protected area, but where the governance authority (i.e. a community, indigenous peoples' group, religious group, private landowner or company) does not wish the area to be reported as a protected area. This could for example be a sacred natural site where nature is highly protected, but the relevant community does not want to draw attention to a place of spiritual importance. In some countries, recognition as a "protected area" by the state brings restrictions, for example on people living within the boundaries. Other examples might include old-growth or other high biodiversity forests permanently set aside in otherwise managed forestry, some community conserved areas or key biodiversity areas managed through regulation or other effective measures whilst still not being fully protected areas.

2   **"Secondary conservation"** refers to active conservation of an area where biodiversity outcomes are not the primary management objective. For example, this might be a conservation corridor, where nature conservation is not a priority, but its existence allows movement of top predators. Some managed forests might fall into this category. Other examples of secondary conservation could include types of traditional agriculture, areas of urban or municipal parks managed for conservation, watershed areas, hunting reserves and long-term fisheries closures. The point (and the distinction from point 3) is that the managers are consciously taking conservation into account when making decisions, even if it is not their first objective.

3   **"Ancillary conservation"** areas delivering *in-situ* conservation as a by-product of management, even though biodiversity conservation is *not* a significant objective. Here conservation is an unplanned benefit from other forms of management: such as some water protection zones, military training grounds or areas set aside between countries for political reasons (demilitarised zones or DMZs), historic marine wreck sites or war graves, sacred sites etc. For example, rich habitat exists between North and South Korea, with somewhat optimistic plans to convert it eventually into a national park (Young-je, 2012).

IUCN has prepared technical guidelines on managing OECMs (IUCN-WCPA Task Force on OECMs, 2019). OECMs are very likely to be included in any future internationally agreed conservation targets and will therefore start to be recorded by countries and thus listed on an additional database linked to the World Database on Protected Areas.

Although the conservation movement was initially very wary of OECMs, in rolling them out they have created a lot of new opportunities alongside a few potential headaches. The implications of this are still being worked out (Jonas, 2018), for example for marine conservation (Laffoley et al., 2017), privately protected areas (Mitchell et al., 2018b) and community conservation (Jonas et al., 2014). The first OECM was officially reported by a government in early 2019: Canadian Forces Base Shilo in south-central Manitoba, a 23,061-hectare military training base that is home to seventeen at-risk species and other wildlife in a near-pristine prairie environment (Government of Canada, 2019). It is likely that more and more OECMs will be recognised as governments start to realise the potential of dual-purpose areas of land and water.

There are however concerns amongst conservationists that OECMs could become an easy option for governments. Rather than establishing protected areas they might identify land and water with vague conservation benefits and make few changes to the existing system. (This also happens with official protected areas, as we'll see in the discussion about effectiveness.) In consequence there has been advocacy to ensure that any future global targets include percentages for both protected areas and OECMs. More positively, OECMs could bring new or existing areas that are important for biodiversity conservation into overall conservation planning and thus help prevent them from being lost or degraded. Most OECMs will not be "created", so much as "recognised"; in other words, existing land and water uses will be formally recognised within the conservation estate. The hope is that by doing so their conservation role will be better safeguarded into the future.

Ancillary conservation in a war grave, Scotland

Scapa Flow is a body of fairly shallow water between several of the Orkney Islands, off the northern coast of Scotland. The islands have been a centre of trade and shipping since long in prehistory. Today, the quiet waters of Scapa Flow have an abnormally large number of sunken ships, some with parts still showing above the water surface. A large part of the German fleet was scuttled there in 1919 to stop them falling into British hands, while during the Second World War HMS Royal Oak was sunk there by a submarine, with the loss of over 800 lives. The wreck of the Royal Oak is now a war grave, as many bodies were never recovered, while the seven German ships that were not salvaged are popular diving spots. A war grave is not a protected area; its primary purpose is to provide peaceful repose for the dead. But the area is fully protected, secure and rich in marine life; it could clearly be a candidate OECM. **ND/SS**

It is fair to say that we still really don't know how OECMs will work out in practice, but they will likely be a significant new tool in the conservation toolbox.

Furthermore, these are not the outer limits of area-based conservation. The definitions of both protected areas and OECMs are actually quite restrictive. OECMs exclude, for example, most commercially managed forests; conservation measures applied to a single species (such as ocean areas with restrictions on take of certain species or groups, like whale sanctuary areas); and any temporary measures such as most fishery closures, agricultural set asides or fallows. Yet all these have some value to nature conservation if used wisely.

There are discussions about more formal recognition of some other types of area-based conservation, including particularly areas that help maintain ecological connectivity (Hilty et al., 2019), which are becoming known as "areas of connectivity conservation". Connectivity is essential to prevent ecological isolation. It helps range expansion, in cases where population has increased above carrying capacity, or when climate has changed and species need to move to re-discover optimal living conditions. Reconnecting previously isolated patches can also help species to re-establish themselves in areas where they previously disappeared. Yet habitat for migration or more regular movement, particularly of larger mammals and birds, can sometimes be very non-natural but still useful in the overall conservation matrix. For example, oil palm plantations contain little in the way of biodiversity but can provide corridors for large cats such as the jaguars (Boron et al., 2016), the Swedish government pays farmers to grow potatoes to feed migrating cranes and we have seen zebra and wildebeest sheltering preferentially in eucalyptus plantation in the buffer zone of iSimangaliso Wetland Park in South Africa (despite the removal of plantations and restoration of savannah within the park). Larger mammals like the cougar will use rivers or pasture (Beier, 1993) while many birds use hedgerows (Haas, 1995), even if they run through areas such as intensive agriculture of urban settlement. Migratory species need stop-over points for resting and feeding, while other species make use of corridors to complete life-cycle movements such as fish ladders traversed by salmon to reach their breeding grounds. Dispersal across large areas is important for long-term population viability, such as when young disperse or in response to population changes; for instance, vegetation patches are used by koalas to disperse through suburban areas in parts of Australia (McAlpine et al., 2006). None of these are protected areas or even necessarily within OECMs but a smart conservation planner would include them in any area-based plans.

Further away still from the conventional picture of a national park or nature reserve, many grazing pastures contain high levels of biodiversity so long as they still maintain a natural species mix (in other words grass has not been replaced with a monoculture of non-native species) and grazing levels are maintained at non-damaging levels. Conservation practice in the Northern Great Plains in the United States and Canada focuses more on encouraging sustainable beef farming as an alternative to conversion to crops than it does on protected areas, building working relationships with ranchers. Similarly, managed forest generally contains less biodiversity than old-growth native forests but may be preferential by far to conversion into soya or oil palm; indeed, well-managed forest with effective poaching control can be better for some wild species than a poorly resourced national park open to uncontrolled depredation.

What appears to be emerging is recognition of a wide sweep of measures: protected areas and OECMs with varying approaches to management, and then areas outside either of these main categories but which nonetheless need to be taken into consideration in planning conservation on a broad scale. However, a major challenge remains in that despite the best efforts of those working on the World Database of Protected Areas and associated databases (Bingham et al., 2019), there are many protected areas still not reported internationally, especially those not managed by governments. Collecting data on OECMs, let alone other types of area-based conservation, presents a major additional challenge and could take decades to achieve.

Our impression is that most conservation planners have not caught up as yet with the new reality or the new options. Many plans still focus entirely on protected areas with the "stricter" management approaches (IUCN Categories I–IV). But this is likely to change over the next few years as more and more governments and NGOs start experimenting with the new options that are becoming available. What this means, or could mean, we'll discuss in the final section.

References

Aerts, R., van Overtveld, K., Haile, M., Hermy, M., Deckers, J., et al. (2006) 'Species composition and diversity of small Afromontane forest fragments in Northern Ethiopia'. *Plant Ecology*, vol. 187, pp 127–142

Beier, P. (1993) 'Determining minimum habitat areas and habitat corridors for cougars'. *Conservation Biology*, vol. 7, pp 94–108

Bingham, H.C., Juffe Bignoli, D., Lewis, E. et al. (2019) 'Sixty years of tracking conservation progress using the world database on protected areas'. *Nature Ecology & Evolution*, vol. 3, pp 737–743. Doi:10.1038/s41559-019-0869-3

Bishop, K., Dudley, N., Phillips, A. and Stolton, S. (2004) *Speaking a Common Language – the Uses and Performance of the IUCN System of Management Categories for Protected Areas*. Cardiff University, IUCN, UNEP and WCMC, Cardiff and Cambridge

Boron, V., Tzanopoulos, J., Gallo, J., Barragan, J., Jaimes-Rodriguez, L., et al. (2016) 'Jaguar densities across human-dominated landscapes in Colombia: The contribution of unprotected areas to long term conservation'. *PLoS One*, vol. 11, no. 5. https://doi.org/10.1371/journal.pone.0153973

Borrini-Feyerabend, G. (2007) *Recognising and Supporting Indigenous and Community Conservation – Ideas and Experience from the Grassroots*. CEESP Briefing Note 9, Cenesta, Tehran

Borrini-Feyerabend, G., Dudley, N., Lassen, B., Pathak, N. and Sandwith, T. (2012) *Governance of Protected Areas: From Understanding to Action*. Best Practice Protected Area Guidelines Series No. 20. IUCN, CBD and GIZ, Gland, Switzerland

Borrini-Feyerabend, G. and Hill, R. (2015) 'Governance for the conservation of nature'. In: Worboys, G.L., Lockwood, M., Kothari, A., Feary, S. and Pulsford, I. (eds.) *Protected Area Governance and Management*. ANU Press, Canberra, pp 169–206

Borrini-Feyerabend, G., Pimbert, M., Farvar, M.T., Kothari, A. and Renard, Y. (2004) *Sharing Power: Learning-by-Doing in Co-Management of Natural Resources Throughout the World*. Earthscan, London

Carlsson, L. and Berkes, F. (2005) 'Co-management: Concepts and methodological implications'. *Journal of Environmental Management*, vol. 75, pp 65–86

CBD. (2018) *Decision Adopted by the Conference of Parties to the Convention on Biological Diversity*. CBD/COP/DEC/14/8, 30 November

Clarke, P. and O'Leary, P. (2018) 'Australia to fund five new indigenous protected areas'. *Pew*, 16 July. www.pewtrusts.org/en/research-and-analysis/articles/2018/07/16/australia-to-fund-five-new-indigenous-protected-areas. Accessed 16 November 2019

Clutton-Brock, T.H. and Ball, M.E. (eds.) (1987) *Rhum: The Natural History of an Island*. Edinburgh University Press, Edinburgh

Dudley, N. (ed.) (2008) *Guidelines for Applying Protected Area Management Categories*. IUCN, Gland, Switzerland

Dudley, N., Higgins-Zogib, L. and Mansourian, S. (2006) *Beyond Belief: Linking Faiths and Protected Areas to Support Biodiversity Conservation*. WWF and Alliance of Religions and Conservation, Gland, Switzerland and Manchester

Government of Canada. (2019) *Canadian Forces Base Shilo and Other Effective Area-Based Conservation Measures*. www.canada.ca/en/environment-climate-change/news/2019/03/canadian-forces-base-shilo-and-other-effective-area-based-conservation-measures.html. Accessed 1 November 2019

Haas, C. (1995) 'Dispersal and use of corridors by birds in wooded patches on an agricultural landscape'. *Conservation Biology*, vol. 9, no. 4, pp 845–854

Hadley, M. (ed.) (2002) *Biosphere Reserves: Special Places for People and Nature*. UNESCO, Paris

Hilty, J., Keeley, A.T.H., Lidicker Jnr, W.Z. and Merenlender, A.M. (2019) *Corridor Ecology*, 2nd edition. Island Press, Covelo, CA

IUCN-WCPA Task Force on OECMs. (2019) *Recognising and Reporting Other Effective Area-Based Conservation Measures*. IUCN, Gland, Switzerland

Jonas, H.D. (ed.) (2018) Special Issue on OECMs *PARKS*, vol. 24.

Jonas, H.D., Barbuto, V., Jonas, H.C., Kothari, A. and Nelson, F. (2014) 'New steps for change: Looking beyond protected areas to consider other effective area-based conservation measures'. *PARKS*, vol. 20, pp 111–128

Khalil, W. and Hani, N. (2014) *Shouf Biosphere Reserve: Field Guide and Information Booklet*. Shouf Biosphere Reserve, Lebanon

Laffoley, D., Dudley, N., MacKinnon, D., MacJinnon, K., Hockings, M. and Woodley, S. (2017) 'An introduction to "other effective area-based conservation measures" under Aichi target 11 of the convention on biological diversity: Origin, interpretation and emerging ocean issues'. *Aquatic Conservation Marine and Freshwater Systems*, vol. 27, no. S1, pp 130–137

Lee, E. (2015) 'Protected areas, country and value – culture tyranny of the IUCN's protected area guidelines for indigenous Australians'. *Antipodes*. Doi:10.1111/anti.12180

Le Thanh Tuyen. (2011) *Coordination Framework and the Planning for Natural Resources Management and Development in the Cat Ba Archipelago Biosphere Reserve, Vietnam*. Paper presented at the workshop, The Cat Ba Archipelago Biosphere Reserve – Towards a Model for Sustainable Development. Cat Ba Biosphere Reserve, 18–19 October

Lopoukhine, N. and de Souza Ferreira Dias, B. (2012) 'Editorial: What does target 11 really mean?' *PARKS*, vol. 18, pp 5–8

Luyoma, J. (1999) *The Hidden Forest: The Biography of an Ecosystem*. Henry Holt and Company, New York

Mallarach, J.M. and Papayannis, T. (2007) *Protected Areas and Spirituality: Proceedings of the First Workshop of the Delos Initiative – Montserrat 2006*. IUCN, Gland, Switzerland

McAlpine, C.A., Brown, M.E., Callaghan, J.G., Lunney, D., Rhodes, J.R., et al. (2006) 'Testing alternative models for the conservation of koalas in fragmented rural-urban landscapes'. *Austral Ecology*, vol. 31, no. 4, p 529

Mitchell, B.A., Fitzsimons, J.A., Stevens, C.M.D. and Wright, D.R. (2018a) 'PPA or OECM? Differentiating between privately protected areas and other effective area-based conservation measures on private land'. *PARKS*, vol. 24, special issue, pp 49–60

Mitchell, B.A., Stolton, S., Bezaury-Creel, J., Bingham, H.C., Cumming, T.L., et al. (2018b) *Guidelines for Privately Protected Areas*. Best Practice Protected Area Guidelines Series No. 29. IUCN, Gland, Switzerland

Monbiot, G. (2017) 'The lake district as a world heritage site? What a disaster that would be'. *The Guardian*, 9 May. www.theguardian.com/commentisfree/2017/may/09/lake-district-world-heritage-site-george-monbiot. Accessed 10 November 2019

Palmer, M. and Finlay, V. (2003) *Faith in Conservation*. The World Bank, Washington, DC

Raz, J. (1988) *The Morality of Freedom*. Clarendon Press, Oxford

Stenke, R. and Chu, X.C. (2004) 'The golden-headed langur (*Trachypithecus poliocephalus poliocephalus*) on Cat Ba Island – status, threat factors, and recovery options'. In: Nadler, T., Streicher, U. and Ha Thang Long (eds.) *Conservation of Primates in Vietnam*. Frankfurt Zoological Society, Hanoi, pp 72–77

Stoll-Kleeman, S., d la Vega-Leinert, A.C. and Schultz, I. (2010) 'The role of community participation in the effectiveness of UNESCO biosphere reserve management: Evidence and reflections from two global surveys'. *Environmental Conservation*, vol. 37, no. 3, pp 227–238

Stolton, S. and Dudley, N. (2008) *Company Reserves: Integrating Biological Reserves Owned and Managed by Commercial Companies into the Global Protected Areas Network – a Review of Options*. WWF International, Gland, Switzerland

Stolton, S. and Dudley, N. (eds.) (2009) *Defining Protected Areas: An International Conference in Almeria, Spain, May 2007*. IUCN, Gland, Switzerland

Stolton, S., Dudley, N. and Beland-Lindahl, K. (1999) 'The role of large companies in forest protection in Sweden'. In: Stolton, S. and Dudley, N. (eds.) *Partnerships for Protection: New Strategies for Planning and Management for Protected Areas*. Earthscan, London

Stolton, S., Redford, K.H. and Dudley, N. (2014) *The Futures of Privately Protected Areas*. IUCN, Gland, Switzerland

Tatarski, M. (2019) *On One Island, a Microcosm of Vietnam's Environmental Challenges, Mongabay*, 22 April. https://news.mongabay.com/2019/04/on-one-island-a-microcosm-of-vietnams-environmental-challenges/. Accessed 19 July 2019

Wilson, J. (2005) 'Removal of grass by scraping to enhance nesting areas for breeding waders at Leighton Moss RSPB reserve, Lancashire, England'. *Conservation Evidence*, vol. 2, pp 60–61

Woodley, S., Bertzky, B., Crawhall, N., Dudley, N., Miranda Londoño, J., et al. (2012) 'Meeting Aichi target 11: What does success look like for protected area systems?' *PARKS*, vol. 18, pp 23–36

Young-je, G. (2012) *Journey to the Ecosystem of the DMZ and the CCL: A Very Special Land*. Korea National Park Service, Seoul

5

DECIDING WHERE PROTECTED AREAS SHOULD BE LOCATED

Ecosystems are interconnected; setting aside areas of land and water in isolation will only be of limited usefulness in conserving biodiversity and other ecosystem services. To stand a serious chance of success, area-based conservation efforts need to be both carefully located to "capture" the most significant ecosystems and species, and to be well enough connected with other ecosystems to allow necessary flow of nutrients, energy and genetic material between different systems. Given that the whole world cannot be one big protected area, this necessarily means incorporating different types of land use into conservation, and the terms landscape approach or seascape approach have come to signify these processes. A wide variety of tools and methods has developed to help identify sites important for biodiversity conservation. Various techniques of systematic conservation planning exist for stitching these together into a coherent mosaic and the developing field of connectivity conservation is important here. These operate at a range of scales, from international plans to maintain pathways for migratory birds to approaches focusing on vaguely defined "landscapes". Understanding how the various forms of area-based conservation fit into these broader approaches is a key step in reaching our ambitious conservation vision.

Everything connects, in ecology as in society: whether or not you subscribe to the theory that the whole earth is acting as a single self-regulating organism (known as the Gaia hypothesis), it is axiomatic that ecosystems do not generally exist in isolation. To be resilient they require regular interaction with other ecosystems. The partial exception of island ecosystems is fascinating to ecologists because if isolated from mainland influences for long enough they have generally evolved unique species, but these ecosystems are also exceptionally fragile to outside change. It is no coincidence that the most spectacular extinctions of recent centuries have occurred, and are occurring, on islands.

The earliest modern protected areas were often set up in a rush to halt the extinction of particular species, without much thought about how they would interconnect; in many countries the overall level of development was still so low that isolation probably didn't appear to be a problem. But protected areas increasingly became islands, cut off from similar ecosystems and additionally depleted through mismanagement or poaching. In the 1980s, this complacency started to be challenged, particularly by Norman Myers in his book *The Sinking Ark*, which argued persuasively that the then current levels of conservation meant that the world was heading for a colossal loss of biodiversity. Norman died while we were writing this book; his thinking changed the trajectory of many conservationists when it first appeared, including our own.

Furthermore, area-based conservation may be the best tool for conservation of biodiversity and ecosystem services, but it is not the only one, nor is it the best in every circumstance. Some conservation challenges do not need dedicated areas set aside from development.

Conservation in Samoa

Due to its long isolation in the tropical Pacific, the island of Samoa has many endemic species, including ten species of birds. Some of these are confined to older forests and do need protected, undisturbed areas to survive, such as the critically endangered tooth-billed pigeon (*Didunculus strigirostris*). However, several others, like the Samoan starling (*Aplonis atrifusca*) and flat-billed kingfisher (*Todirhamphus recurvirostris*), are long-adapted to living among humans and exist happily in gardens and other cultural landscapes. For these birds, conservation priorities are more about actions to control potentially invasive species that could upset the delicate balance of the ecosystem, like the brown tree snakes (*Boiga irregularis*) that have decimated populations of native bird species on the island of Guam. Effective conservation planning involves stepping back and looking at the broad picture, all the way from landscape/seascape to conservation implications beyond national borders and considering a range of management options, which will likely centre largely on various forms of area-based conservation, but not be confined to these. **ND/SS**

Where to focus attention

Over the past twenty years there has been a huge effort, and some spectacular institutional and personal disputes, about different methods of identifying the most important sites on which to focus conservation. Some methods concentrate most on areas at risk for instance, some on overall biodiversity richness, others on uniqueness (such as endemism) and subsequent vulnerability, and some on various measures of intactness. Most have originally been drawn up, or at least had the criteria defined, by fairly small groups of people, overwhelmingly in wealthy countries such as the United States, and this bias also comes through in the selection process. Furthermore, many are based on a mapping process often made with inadequate data and little reality checking of what is actually happening "on the ground".

Underlying many of these systems is a focus on where species are at risk. The IUCN Red List databases, organised and populated by its Species Survival Commission, form the framework around which most conservation prioritisation assessments are set. Red Lists exist at global and national scale and individual reports are brought together meticulously, usually by volunteers, and often for apparently obscure groups. The IUCN Red List of Threatened Species™ is essentially a checklist of taxa that have undergone an extinction risk assessment using the IUCN Red List Categories and Criteria. It divides species into nine categories: Not Evaluated, Data Deficient, Least Concern, Near Threatened, Vulnerable, Endangered, Critically Endangered, Extinct in the Wild and Extinct. Assessments for species, groups of species or countries are proposed by experts, following an agreed set of criteria, and vetted for approval by specialists appointed by IUCN. Currently there are well over 100,000 species on the Red List, with more than 28,000 species assessed as being threatened with extinction, including 40 per cent of amphibians, 34 per cent of conifers, 33 per cent of reef building corals, 25 per cent of mammals and 14 per cent of birds. While this still represents a fraction of the world's total species, it provides increasingly comprehensive coverage for species such as mammals, birds and some groups of higher plants. More recently, IUCN has also drawn up a parallel Red List of Ecosystems (Rodríguez et al., 2015), which is compiling lists of ecosystems at risk and will play a similar role in informing conservation prioritisation exercises. Plans are also well advanced for parallel green lists for species (Akçakaya et al., 2019) and ecosystems, to provide a way of measuring recovery.

Early attempts at identifying the most important places to focus conservation, such as the Centres of Plant Diversity study by WWF and partners (Davis et al., 1994) and the Valuable Grassland Area analysis for the Southern Cone of Latin America (Bilenca and Miñarro, 2004), relied on expert opinion to identify priority sites. These should not be dismissed; pooling knowledge from a group of experts is certainly the quickest and far from the least accurate way of providing a broad overview.

More systematic methods generally start by classifying the planet according to similarities and differences in geography and ecology. At the top are the so-called biogeographic realms or ecozones, eight broad areas defined largely by their plant assemblages and first proposed by the Hungarian ecologist Miklos Udvardy (1975). Underneath these, at least in a classification suggested by WWF, come "bioregions" and then smaller still are "ecoregions". WWF distinguished 867 terrestrial ecoregions in a major research effort that also synthesised many previous attempts to define ecoregions (Olson et al., 2001). By no means all ecologists agree with the WWF system and continued debate is almost inevitable given that we are imposing human constructs and hard borders on complicated and evolving natural processes.

Nonetheless, the ecoregions have been widely used as planning instruments. Slightly later, WWF US identified a *Global 200* list of (actually 238) terrestrial, freshwater and marine ecoregions which WWF experts believed were most crucial to the conservation of biodiversity (Olson and Dinerstein, 2002). So, for example, the Queensland Tropical Rainforests ecoregion covers 32,700 km^2 of north-eastern Australia although remaining rainforests are in relatively small patches. The ecoregion is in the Global 200 partly because it is judged to be the world's best living record of major stages in the evolutionary history of land plants and also contains unique species such as the flightless southern cassowary (*Casuarius casuarius*) bird. The Global 200 was a major focus of the organisation's work for a decade although now seems to be talked about less frequently.

At around the same time, Conservation International, drawing on earlier work by Norman Myers, recognised thirty-six biodiversity hotpots, biogeographic regions with significant biodiversity threatened by humans, covering 2.4 per cent of the Earth's surface but with over half the world's plant species as endemics. To qualify as a hotspot a region must have at least 1,500 endemic vascular plants and 30 per cent or less of its original vegetation; in other words, be both rich and threatened from a botanical viewpoint (Myers et al., 2000). Examples include the Western Ghats in southern India and Sri Lanka and the Mountains of Central Asia. This analysis was more consciously policy-orientated than the original classification of ecoregions, although it had parallels with the Global 200. A major policy realignment by Conservation International has reduced attention on the hotspots although the Critical Ecosystem Partnership Fund continues to support conservation in many of these hotspots.

BirdLife International identified global lists of Endemic Bird Areas (Stattersfield et al., 1998) and, more comprehensively, over 12,000 Important Bird Areas (IBA), judged critical for maintaining the world's bird species; when complete it is expected to include around 15,000 sites covering 7 per cent of the world's land surface. To be listed as an IBA, a site must contain critically endangered or endangered bird species according to the IUCN Red List, or restricted range or biome-restricted species or important congregations of species. Although originally developed for terrestrial sites, some 3,000 coastal and high seas IBAs have also been identified. This is undoubtedly the most comprehensive global classification to date, with information from all continents; something of this detail is probably only possible for birds at the present time.

An important bird area in Wales

The Dyfi estuary in mid Wales where we live is an IBA, listed in 2007 as a site for species threatened at a European level (which means it is not a "global IBA"), being one of the five most important sites for a species or sub-species threatened in the region. Presumably the listing refers to the over-wintering of the Greenland white-fronted goose (*Anser albifrons flavirostris*), a distinct race of the goose that breeds in Greenland, visits Iceland as a passage migrant and over-winters in Ireland and the UK. The IBA database lists the site as in near-favourable condition, which is probably correct, because most of the area is protected either in a national nature reserve or a reserve owned and managed by the Royal Society for the Protection of Birds (although some duck shooting is permitted in the former). However, for reasons not yet fully understood the number of white-fronted geese visiting the estuary has declined dramatically in the last few years, from a few hundred down to around a dozen, so the values for which the IBA was listed have probably declined. (The site is important for many other bird species, but not of a rarity to be the focus of the IBA.) **ND/SS**

The fact that a site is rich in birds does not necessarily mean that the same is true for other plant and animal groups. Plantlife has identified Important Plant Areas for several countries and found considerable divergence from IBAs (Anderson, 2002). The Alliance for Zero Extinction lists sites around the world that are the only known location for a particular endangered species (Funk et al., 2017). The Wildlife Conservation Society has focused instead on intactness as a measure of importance and has drawn up a list of places that qualify as "Last of the Wild" using an overlay method to identify the Human Influence Index (Sanderson et al., 2002). Greenpeace International developed criteria for intact forest landscapes in 2006; this effort has since involved many more organisations and includes natural forests of at least 50,000 hectares, at least 10 km wide at the broadest place and at least 2 km wide in corridors (Potapov, 2008). There are still other systems.

More recently, IUCN and partners have made a huge effort to standardise prioritisation, drawing up criteria and indicators for Key Biodiversity Areas (KBAs) (IUCN, 2016). A KBA site needs to meet one or more criteria, covering threatened biodiversity, geographically restricted biodiversity, ecological integrity, biological processes and irreplaceability by quantitative analysis. Each of these has several subsections and, importantly, quantified thresholds. So, for example under section D3 ("Biological processes – recruitment sources") a site would be recognised as a KBA if it predictably produces propagules (seeds, spores etc) or larvae of juveniles that maintain at least 10 per cent of a species' global population. For instance, a spawning ground for a particular fish species that regularly produces over a tenth of its world population would be a KBA. This is the most rigorous prioritisation system to date and KBAs are being referred to in international policy. However, the current weakness of KBAs is that most sites listed on the database at the moment are BirdLife's IBAs, and bird diversity is not always a surrogate for diversity in other species. More importantly, IBAs were not set to exactly the same criteria KBAs and their listing is therefore provisional. A comprehensive and accurate global list of sites still seems to be decades away.

Nevertheless, analyses are emerging, including in some apparently unlikely places. The NGO Nature Iraq produced a nationwide survey of KBAs, building initially on a 1994 desk analysis that identified forty-two IBAs. Six years of fieldwork throughout the country took place from 2005–2010, often in difficult conditions, and was followed by a period of analysis. Nature Iraq recognized eighty-two KBAs based on a wide range of taxa, including many endemic plants and covering mountain, desert and wetland ecosystems, with results jointly published with the Ministry of Health and Environment to give an important level of official endorsement (Bachmann et al., 2017).

Huge gaps remain in our global understanding of species and ecosystems, and critics complain that many key decisions on land and water use in hitherto less developed areas will have long been decided before all KBAs have been identified. These global analyses also leave open the question of what conservation should take place in areas *outside* the priority sites; most of Europe does not feature at all in many global conservation prioritisation exercises due to its long habitation and the extent to which ecosystems have been modified (for example there were just two Global 200 ecoregions from western Europe out of the 237 listed). This isn't surprising, as the Global 200 exercise was in large part an effort to channel a large proportion of available conservation dollars into high risk sites in the tropics and away from western Europe, where conservation was already quite advanced. But it does lead to some anomalies. Do those of us who live in the continent simply stop bothering?

Furthermore, all of these broad-brush systems have been developed by non-governmental organisations, and the need for publicity and fund-raising has influenced both the variety of approaches and the vigour with which they have been debated and compared. All have their strengths, weaknesses and biases. Have they really made a difference to where the conservation dollar is invested? To some extent yes, but perhaps less than the originators imagined. We will return to this in discussion about effectiveness in the next chapter.

Landscape- and seascape-level analyses down to the site

These global analyses are useful at identifying broad areas that need attention but may not always help make decisions on a local scale or within a site. The KBA Partnership is clear that not all KBAs or the whole territory of a KBA necessarily needs to be in a protected area so long as its core values are maintained; KBAs and other similar designations are not blanket descriptors but will often contain areas of greater or lesser importance from a conservation perspective. Additionally, many areas that fall outside global-scale ranking exercises will contain important species and ecosystems. More local scale assessment systems are available to help make selections within a site.

One of the best-known site-level tools is the High Conservation Value Areas (HCVA) concept (Brown et al., 2013), additionally developed for forests but later extended to freshwater (Abell et al., 2015) and grasslands. HCVAs are defined according to six conservation "values", each with a series of indicators, three relating to broadly ecological values and three to socio-economic and cultural values:

HCV 1 Species diversity: concentrations of biodiversity including endemic species, and rare, threatened or endangered species, significant at global, regional or national levels.

HCV 2 Landscape-level ecosystems and mosaics: large ecosystems and ecosystem mosaics, significant at global, regional or national levels, containing viable populations of almost all naturally occurring species in natural patterns of distribution and abundance.

HCV 3 Ecosystems and habitats: rare, threatened or endangered ecosystems, habitats or refugia.

HCV 4 Ecosystem services: basic ecosystem services in critical situations, including protection of water catchments and control of erosion of vulnerable soils and slopes.

HCV 5 Community needs: sites and resources fundamental for satisfying the basic necessities of local communities or indigenous peoples (for livelihoods, health, nutrition, water etc), identified through engagement with these peoples.

HCV 6 Cultural values: sites, resources, habitats and landscapes of global or national cultural, archaeological or historical significance, and/or of critical cultural, ecological, economic or sacred importance for the traditional cultures of local communities or indigenous peoples, identified through engagement with these peoples.

Site-level assessments such as HCVA are not isolated from global assessments and for example will often draw on information about presence of KBAs, Red Lists or similar data sets in their analytical framework but also including within the analysis far more local-scale indicators. Similarly, most global assessments are made up of innumerable site-level data. It will be noted that HCVAs, like many more localised analyses, consciously include elements important for human society alongside "pure" conservation information. This is important; we know a lot about ecosystem services at a site level, but global scale analyses are still in their infancy, although this situation is changing fast.

The importance of local knowledge

Little discussed, in no-one's toolbox and not the usual subject of peer-reviewed papers, local experience is perhaps one of the most underrated factors explaining why protected areas are established. But perhaps it is one we need to take more note off. Many privately protected areas are set up either by individuals or NGOs because someone knows that a rare species resides in a small plot of land or in a specific body of water, or a remnant of forest hangs on in a small and rarely visited valley, or an economically unimportant island has a magnificent sea-grass bed. We recently visited a privately protected area along with the passionate founder of the organisation that manages it. We were high in the Andes in an area heavily grazed and often inhospitably wet and windy. It reminded us of Wales, although our lack of breath made it clear how much higher we were. We were there to see a forest restoration project. As we walked along a stony path, with alpacas grazing in the distance and snow-covered mountain peaks above such a forest seemed unlikely. But suddenly we turned a corner and on the sheltered side of a valley there was a completely unexpected forest of native *Polylepis* trees and a reforested area of 10,000 trees planted by local people. Constantino took great pleasure in telling us how scientists told him that *Polylepis* reforestation was next to impossible at such altitude (up to 4,400 metres). But local knowledge told a different story, and led to both protection of the forest and local commitment to plant and tend the young trees. And this is a story that is far from unique. All around the world people are protecting places and species that are special to them. A moment after finishing this section I read of the heartening discovery of populations of silver-backed chevrotain or two-tone mouse deer (*Tragulus versicolor*) in Vietnam. Having not been seen for nearly thirty years the deer was caught in camera traps placed in areas identified by local people. While global overviews can take decades, a lot of knowledge is already out there if we only know who to ask. **SS**

From the perspective of planning area-based conservation, it is not enough to know where biodiversity is found, but whether or not it is existing in conditions where it is likely to survive. One way of getting some of this information is to see whether or not viable populations exist in protected areas or other forms of area-based conservation. The methodology of gap analysis has been developed to help.

Gap analysis

In a conservation context, gap analysis is a method to identify biodiversity not adequately conserved within a protected area network or through other effective and long-term conservation measures. It was developed around twenty-five years ago and was an important element in the CBD's Programme of Work on Protected Areas (Dudley and Parrish, 2006), which encouraged the development of many national studies.

Gap analysis is usually applied to large areas, ideally to the whole of an ecologically defined region such as an ecoregion, although in practice it is often done for countries, or sub-national areas. Gap analysis can vary from a simple exercise based around comparison of species with existing protected area networks to complex studies that require data gathering, mapping and software decision packages to determine optimal protected area networks. However complicated the process used, gap analyses start by identifying the species or ecosystems that are most important to consider (the focal biodiversity) and deciding on targets for their conservation, such as minimum safe population size or area of ecosystem. The occurrence of these focal species is then mapped and compared with existing protected areas to identify gaps in the system. Some prioritisation is usually needed to identify the most important gaps to be filled. In most cases this "ideal" protected area network then needs a dose of reality by judgements about what is politically and socially possible, a process that can take years of negotiation, before settling on a plan.

Gaps in protected area networks come in several forms, which can be divided for convenience into:

Representation gaps: there are either (1) no representations of a species or ecosystem in any protected area, or (2) there are not enough examples of the species/ecosystem represented to ensure long-term protection.

Ecological gaps: while the species/ecosystem is represented in the protected area system, the occurrence is either of inadequate ecological condition, or the protected area(s) fail to address the movements or specific conditions necessary for the long-term species survival or ecosystem functioning.

Management gaps: protected areas exist, but management regimes (management objectives, governance types or management effectiveness) do not provide full security for all the species or ecosystems given the local conditions.

Some important principles of gap analysis include to:

- **Ensure full representation** across biological scales (species and ecosystems) and biological realms (terrestrial, freshwater and marine);
- **Aim for redundancy** of examples of species and ecosystems within a protected area network to capture genetic variation and protect against unexpected losses;
- **Design for resilience** to ensure protected area systems withstand stresses and changes, such as climate change;
- **Consider "representation"** gaps, **"ecological"** gaps and **"management"** gaps in the analysis;
- **Employ a participatory approach,** collaborating with key stakeholders in making decisions about protected areas; and
- **Make protected areas system design an iterative process** in which the gap analysis is reviewed and improved as knowledge grows and environmental conditions change.

Identified gaps do not always mean setting up new protected areas; conservation planners have several options. Expanding the size of an existing protected area is one option, either by enlarging the borders of the protected area or by agreeing a variety of sympathetic management practices in buffer zones (e.g. controlled game zones, sustainable agriculture or areas of controlled fishing), which might become OECMs. Changing the shape of the protected area can sometimes help, for instance by bringing spawning areas for fish under protection in a marine protected area or bringing a fragment of remaining natural forest. Finally, changing the governance arrangement, for instance from management by a government agency to a co-management arrangement or to management by local people can also improve conservation success.

Systematic conservation planning

Standardised approaches to planning at broader landscape or ecoregional scales have emerged under the general name of systematic conservation planning (Margules and Pressey, 2000). Some have been criticised as being top down, where all major decisions are taken by a minority before anyone else gets involved – indeed early models deliberately aimed to set a conservation vision first without considering social or economic consequences and tried to retrofit these into the plan at a later date. Their widespread failure was completely predictable. More recently, models have emerged that recognise the need for a broad level of public and institutional consensus if plans are to be more than theoretical patterns on a map. Planners have looked at integration with other approaches, such as Key Biodiversity Areas (Smith et al., 2018). There are lots of variations; one ten-stage process is described in the following.

- Stage 1. **Identifying broad goals.** Developing a broad, idealized 50–100-year goal or vision for conservation within the region, in collaboration with other landowners, resource users and interested rightsholders and stakeholders.
- Stage 2. **Identifying data needs, assembling and evaluating data.** Working with the collaborators identified earlier to agree on the information needed, then to collect and analyse relevant data. This will include data on biodiversity and ecosystem services and also on issues like economic development, needs and wants of local communities and human-wildlife conflict, along with threats to ecological and human values. Indictors will be needed to avoid a huge research job: identifying species or habitats that can act as "surrogates" or representatives for overall biodiversity and critical social indicators that give a good overview of human wellbeing. New data should be collected if needed.
- Stage 3. **Formulating targets.** Including quantitative conservation targets for species, vegetation types and for minimum size, connectivity and other design criteria for any area-based conservation. Qualitative targets can also be agreed at this time (e.g. favouring the most intact areas of vegetation). There are also likely to be targets for ecosystem services, particular social benefits from the protected area and targets such as minimisation of human-wildlife conflict.
- Stage 4. **Reviewing the effectiveness of existing protected and conserved areas.** Measuring the extent to which the agreed quantitative targets have already been met by the existing network of area-based conservation sites (e.g. protected areas and OECMs) and assessing the effectiveness of the network in excluding threatening processes and facilitating management. This may require a gap analysis.
- Stage 5. **Selecting additional protected areas and OECMs.** Using the existing network of area-based conservation sites as a basis, identifying potential new protected areas or OECMs. This process can be helped by using decision support software but needs to involve all stakeholders and rightsholders

and to take full account of the need for Free Prior and Informed Consent (FPIC) of indigenous people and other local people.

- Stage 6. **Identifying additional sites of nature conservation value**. Filling remaining conservation gaps by considering the potential role of managed landscapes including farmland, forestry, freshwaters and marine fishing areas, identifying what if anything needs to change in order for these sites to play an effective role in nature conservation. This stage involves working with rights-holders, owners and managers about the potential for incorporating such sites within a conservation plan.

- Stage 7. **Implementing conservation action on the ground**. At this stage final decisions need to be taken about whether particular sites should be protected areas or OECMs. If the former, what type of management and governance is most appropriate; for example a strictly protected area or a protected landscape with traditional management taking place alongside conservation?

- Stage 8. **Agreeing any compensation packages**. If people have agreed to forgo some benefits in favour of conservation (access rights, resource user rights etc), suitable compensation packages will often need to be negotiated and agreed by all parties.

- Stage 9. **Maintaining and monitoring the network**. Setting conservation goals in individual protected areas and OECMS, implementing management actions and monitoring key indicators of success, with modifications to management as necessary. Monitoring needs to take place across the whole landscape and include consideration of the role of managed areas as well as the area-based conservation network.

- Stage 10. **Adaptive management**. Noting the condition of target species and habitats and changing management if trends are deteriorating. Nothing works perfectly and most conservation plans are a compromise between the needs of biodiversity and of other stakeholders. The status of conservation targets identified in stage 3 gives a quick reference, but managers need to look also at overall conditions within the area. Things will change: land ownership for instance and attitudes of other people in the landscape or seascape, human demographic changes will bring challenges or opportunities, climate change will probably alter things and new funding streams may become available. A plan is not a static thing set in stone, but something that evolves over time.

But protected areas are still often put in the wrong places

Unfortunately, despite a plethora of tools, many protected areas are inadequately planned, isolated, too small, missing key components or simply in the wrong place, as we will see in the section on effectiveness. As a result, the world's protected areas contain a biased and incomplete sample of biodiversity, and there is a need for a more systematic approach to planning. After the broad sweep analysis, with its generalisations and approximations, local-level planning is an altogether messier affair involving not only what conservation biologists think is important but the needs, wants, prejudices and opinions of a mass of other stakeholders and rightsholders. Area-based conservation networks need to be carefully designed if they are to be effective at conserving biodiversity, but outside the neat offices of academic research institutions, planning is seldom a simple or logical affair, and almost every real-life conservation plan is instead to some extent or other a compromise. This doesn't mean that the process of planning is not critically important in achieving success.

References

Abell, R., Morgan, S.K. and Morgan, A.J. (2015) 'Taking high conservation value from forests to freshwaters'. *Environmental Management*, vol. 56, no. 1, pp 1–10

Akçakaya, H.R., Bennett, E.L., Brooks, T.M., Grace, M.K., Heath, A., et al. (2019) 'Quantifying species recovery and conservation success to develop an IUCN green list of species'. *Conservation Biology*, vol. 32, no. 5, pp 1128–1138

Anderson, S. (2002) *Identifying Important Plant Areas in Europe: A Site Selection Manuel for Compilers*. PlantLife, London

Bachmann, A., Alwash, A. and Lami, A. (eds.) (2017) *Nature Iraq (2017): Key Biodiversity Areas of Iraq: Priority Sites for Conservation and Protection*. Tablet House Publishing, Huntington Beach, CA

Bilenca, D. and Miñarro, F. (2004) *Identifiación de Areas Valiosas de Pastizal (AVPs) en las pampas y campos de Argentina, Uruguay y Sur de Brasil*. Fundacion Silvestre, Buenis Aires, Argentina

Brown, E., Dudley, N., Lindhe, A., Muhtaman, D.R., Stewart, C. and Synnott, T. (eds.) (2013) *Common Guidance for the Identification of High Conservation Values*. HCV Resource Network, Oxford

Davis, S.D., Heywood, V.H. and Hamilton, A.C. (1994) *Centres of Plant Diversity: A Guide and Strategy for Their Conservation*, 3 volumes, IUCN, Cambridge and WWF, Gland, Switzerland

Dudley, N. and Parrish, J. (2006) *Closing the Gap: Creating Ecologically Representative Protected Area Systems*. CBD Technical Series 24. Convention on Biological Diversity, Montreal

Funk, S.M., Conde, D., Lamoreux, J. and Fa, J.E. (2017) 'Meeting the Aichi targets: Pushing for zero extinction conservation'. *Ambio*, vol. 46, no. 4, pp 443–455

IUCN. (2016) *A Global Standard for the Identification of Key Biodiversity Areas*, Version 1.0. IUCN, Gland, Switzerland

Margules, C.R. and Pressey, R.L. (2000) 'Systematic conservation planning'. *Nature*, vol. 243, p 253

Myers, N., Mittermeier, R.A., Mittermeier, C.G., da Fonseca, G.A.B. and Kent, J. (2000) 'Biodiversity hotspots for conservation priorities'. *Nature*, vol. 403, pp 853–858

Olson, D.M. and Dinerstein, E. (2002) 'The global 200: Priority ecoregions for global conservation'. *Annals of the Missouri Botanic Gardens*, vol. 89, pp 199–224

Olson, D.M., Dinerstein, E., Wikramanayake, E.D., Burgess, N.D., Powell, G.V.N., et al. (2001) 'Terrestrial ecoregions of the world: A new map of life on earth'. *BioScience*, vol. 51, no. 11, pp 933–938

Potapov, P., Yaroshenko, A., Turubanova, S., Dubinin, M., Laestadius, L., et al. (2008) 'Mapping the world's intact forest landscapes by remote sensing'. *Ecology and Society*, vol. 13, p 51. www.ecologyandsociety.org/vol13/iss2/art51/

Rodríguez, J.P., Keith, D.A., Rodríguez-Clark, K.M., Murray, N.J., Nicholson, E., et al. (2015) 'A practical guide to the application of the IUCN red list of ecosystems criteria'. *Philosophical Transactions of the Royal Society B*, vol. 370, Article Id. 20140003

Sanderson, E.W., Jaiteh, M., Levy, M.A., Redford, K.H., Wannebo, A.V., et al. (2002) 'The human footprint and the last of the wild'. *BioSciences*, vol. 52, no. 10, pp 891–904

Smith, R.J., Bennun, L., Brooks, T.M., Butchart, S.M., Cuttelod, A., et al. (2018) 'Synergies between key biodiversity areas and systematic conservation planning approaches'. *Conservation Letters*, p e12625

Stattersfield, A.J., Crosby, M.J., Long, A.J. and Wege, D.C. (1998) *Endemic Bird Areas of the World: Priorities for Biodiversity Conservation*. BirdLife International, Cambridge

Udvardy, M.D.F. (1975) *A Classification of the Biogeographical Provinces of the World*. IUCN Occasional Paper No. 18. IUCN, Morges, Switzerland

6

SETTING TARGETS FOR CONSERVATION

The international community was slow to set targets for protected areas although this has since changed. Milestones include the CBD's Programme of Work on Protected Areas in 2004, the setting of global percentage targets for terrestrial and marine protected areas by the CBD in 2010 and the adoption of these by the Sustainable Development Goals in 2015. At a regional level, the Natura 2000 process in the European Union is the primary example of a continent-wide strategy with government backing. The long-term success or failure of these policy initiatives will be heavily influenced by decisions made over the next few years, making this a critical time for area-based conservation.

The world took a long time to get around to thinking comprehensively about protected area systems, let alone other forms of area-based conservation. IUCN began to talk about 10 per cent protection in the 1980s, through the precursor body to today's World Commission on Protected Areas (WCPA), but very much as a theoretical exercise.

The three global conventions agreed in Rio de Janeiro in 1994 at the so-called Earth Summit were aimed at climate, biodiversity and desertification. The Convention on Biological Diversity was the closest match to the ideas being discussed here, but in its early years focused more on biodiversity as intellectual property than biodiversity conservation as such. This gradually changed and following overtures from WCPA began to look more closely at conservation mechanisms. Its Programme of Work on Protected Areas (POWPA) was agreed in 2004, a detailed and ambitious plan that drew heavily on the outcomes of IUCN's fifth World Parks Congress, held in Durban, South Africa, the previous year (CBD, 2004). The programme, which was to run from 2004 to 2015, had ambitious goals clustered around site selection and expansion of the system, increasing governance quality, participation and equity, a range of enabling activities and standards, assessment and monitoring (Dudley et al., 2005).

Some countries took the POWPA very seriously and it helped to promote activities such as national gap analyses. But after six years, it was superseded (perhaps swamped would be more accurate) by the 20 Aichi Biodiversity Targets. Aichi Target 11 focused on protected areas and reads:

> By 2020, at least 17 per cent of terrestrial and inland water areas and 10 per cent of coastal and marine areas, especially areas of particular importance for biodiversity and ecosystem services, are conserved through effectively and equitably managed, ecologically representative and well-connected systems of protected areas, and other effective area-based conservation measures, and integrated into wider landscape and seascape.

The target fitted a lot into one sentence. If we ignore the 17 per cent, a result of horse trading at 3 o'clock in the morning, the target encapsulated everything that conservationists should be thinking about – representation, connectivity, social equitability, effective management – for protected areas but also pushed the debate further by mentioning "other effective area-based conservation measures". As we've described, that started a debate about how area-based conservation might extend beyond protected areas and may yet be the most significant legacy of the whole Aichi target-setting process.

The Aichi goals for terrestrial and marine areas were repeated almost verbatim in the Sustainable Development Goals (goals 14 and 15), with a 2020 deadline and an understanding that these SDG goals will be revised after that date depending on what the CBD decides the next round of targets should be.

The Aichi targets are generally acknowledged to have failed. Aichi 11 is an exception; it is likely that the two numerical targets will be achieved, or virtually achieved, by the target date – a rare success for international targets. Whether or not sub-targets referring to ecological representation, quality of management effectiveness, equity and so on have been as successful is currently a matter of debate. Serious mismatches between the location of protected areas and of threatened species have been identified (Venter et al., 2017). Analysis in 2013 found that only 22 per cent of Important Bird Areas were completely covered by protected areas for example, with protection of Key Biodiversity Areas actually slowing down (Juffe-Bignoli et al., 2014). In an editorial essay published in *PARKS* (Gannon et al., 2017) there was a generally upbeat assessment from the CBD secretariat and partners. Critics claim that percentage targets for protected areas can act as a perverse incentive by simply encouraging governments to set aside unimportant areas and could actually undermine broader conservation goals (Visconti et al., 2019), a proposal that created some kickback, including from us (Woodley et al., 2019b). Our own opinion is that Aichi 11 has been critical in building understanding and support for both area-based conservation and the associated implications, but that it currently does not go far enough and the next few years will be critical in determining whether it is the start of a more profound change in management of the planet's terrestrial and marine ecosystems, or a blip that will have far fewer long-term consequences.

This is not to say that protected areas are all in the right place or all being designated with the correct management regimes (Jantke et al., 2018). The sudden rush to create marine protected areas for instance has ended up with some very large MPAs in places with little pressure, far less protection in coastal waters and some designations that are so weak that they will clearly not protect either fish stocks or wider marine biodiversity. There is a rich literature available on the need for greater specificity in the location of area-based conservation (e.g., Venter et al., 2014) and it is to be hoped that future conservation efforts are far more strategic.

The other two Rio Conventions do not have specific goals for area-based conservation although they do imply the need for such actions in several of their programmes and targets. The UN Framework Convention on Climate Change (UNFCCC) has, after a long period of inaction on this issue, given higher profile to the need to address greenhouse gas emissions from terrestrial and aquatic ecosystems. The various REDD (Reduced Emissions from Deforestation and forest Degradation) funding packages, along with an array of voluntary carbon market initiatives and the Green Climate Fund, also provide financial incentives for management that secures carbon in vegetation and soils. The UN Convention to Combat Desertification (UNCCD) has a target of Land Degradation Neutrality and recognises that protected areas and other forms of area-based conservation will be an important part of any strategy to achieve LDN (UNCCD, 2017).

Regional target setting processes are far less common. In Europe, The Natura 2000 programme aims to establish a set of conservation areas that protect vulnerable or endangered species throughout the continent. Established in 1992, it designates a series of Special Areas of Conservation (SACs) and Special Protection Areas (SPAs) designated respectively under the Habitats Directive and the Birds Directive. EU members differ in the way they approach Natura 2000; some countries designate all N2000 areas as formal protected areas, while others use a wider portfolio of conservation strategies. Currently over 27,000 sites have been designated covering around 18 per cent of the EU territory; the quality of conservation varies, but this remains one of the most ambitious international conservation efforts to date.

Debates about the future are ongoing. In the marine realm, IUCN members agreed at the 2016 World Conservation Congress that 30 per cent of the ocean should be in a marine protected area, an ambitious ramping up of current protection levels. A recent study of 1,656 papers focusing on areas necessary to be conserved to preserve global biodiversity varied in their estimation of the proportion of the terrestrial area required for conservation from 30 up to 80 per cent. Proposals for "30 per cent by 2030" are emerging, for example from NGOs like WWF, while a wider set of objectives aiming towards protection of half the planet are being backed by luminaries such as E.O. Wilson (Wilson, 2016). The "global deal for nature" concept suggests 30 per cent protection with another 20 per cent set aside for climate stabilisation areas by 2030 and identifies 67 per cent of natural ecoregions that could meet the 30 per cent protection level (Dinerstein et al., 2019). The need for greater protection of ocean areas, as opposed to just coastal marine areas, is also getting higher profile (e.g. Freestone et al., 2017).

As noted earlier, numerical targets of this kind are controversial. In the marine context, for example, critics argue that they can persuade governments to promote the designation of MPAs in irrelevant places or create MPAs with so many exemptions in terms of protection that they make no difference to conservation (Agardy et al., 2016). It has been noted that several governments have set "marine protected areas" that only ban activities that do not occur anyway, allowing ongoing types of extraction to continue under a patina of "conservation", a classic example of greenwashing. Supporters of area-based targets recognise the dangers and note that they should not be separated from a range of quality criteria.

The world is currently gearing up to re-set the Aichi targets with a 2030 deadline; as we write it is still very difficult to tell which way the discussions will go (Woodley et al., 2019a). Some of the implications of this are discussed in the final part of the book.

References

Agardy, T., Claudet, J. and Day, J.C. (2016) '"Dangerous targets" revisited: Old dangers in new contexts plague marine protected areas'. *Aquatic Conservation: Marine and Freshwater Ecosystems*, vol. 26, no. S2

CBD. (2004) *Programme of Work on Protected Areas*. UNEP and the Secretariat of the Convention on Biological Diversity, Nairobi and Montreal.

Dinerstein, E., Vynne, C., Sala, E., Joshi, A.R., Fernando, S., et al. (2019) 'A global deal for nature: Guiding principles, milestones, and targets'. *Science Advances*, vol. 5, no. 4, p eaaw2869

Dudley, N., Mulongoy, K.J., Cohen, S., Stolton, S., Barber, C.V., et al. (2005) *Towards Effective Protected Area Systems: An Action Guide to Implement the Convention on Biological Diversity Programme of Work on Protected Areas*. Secretariat of the Convention on Biological Diversity, Montreal, Technical Series No. 18

Freestone, D., Laffoley, D., Douvere, F. and Badman, T. (2017) *World Heritage in the High Sea: An Idea Whose Time Has Come*. World Heritage Reports No. 44. UNESCO, Paris

Gannon, P., Seyoum-Edjigu, E., Cooper, D., Sandwith, T., Ferreira de Souza Dias, B., et al. (2017) 'Status and prospects for achieving Aichi biodiversity target 11: Implications of national commitments and priority actions'. *PARKS*, vol. 23, no. 2, pp 13–26

Jantke, K., Jones, K.R., Allan, J.R., Chauvenet, A.L.M., Watson, J.E.M., et al. (2018) 'Poor ecological representation by an expensive reserve system: Evaluating 35 years of marine protected area expansion'. *Conservation Letters*, vol. 11, no. 6. Doi:10.1111/conl.12584

Juffe-Bignoli, D., Burgess, N.D., Bingham, H., Belle, E.M.S., de Lima, S.G., et al. (2014) *Protected Planet Report 2014*. UNEP World Conservation Monitoring Centre, Cambridge

UNCCD (2017) Global Land Outlook. UN Convention to Combat Desertification, Bonn, Germany.

Venter, O., Fuller, R.A., Segan, D.B., Carwardine, J., Brooks, T., et al. (2014) 'Targeting global protected area expansion for imperilled biodiversity'. *PLoS Biology*, vol. 12, no. 6, p e1001891

Venter, O., Magrach, A., Outram, N., Klein, C.J., Possingham, H.P., et al. (2017) 'Bias in protected area location and its effects on long-term aspirations of biodiversity conventions'. *Conservation Biology*, vol. 32, no. 1, pp 127–134

Visconti, P., Butchart, S.H.M., Brooks, T.M., Langhammer, P.F., Marnewick, D., et al. (2019) 'Protected area targets post- 2020'. *Science*, vol. 364, no. 6437, pp 239–241

Wilson, E.O. (2016) *Half-Earth: Our Planet's Fight for Life*. W.W. Norton & Company, New York

Woodley, S., Baille, J.E.M., Dudley, N., Hockings, M., Kingston, N., et al. (2019a) 'A bold successor to Aichi target 11'. *Science*, vol. 365, no. 6454, pp 649–650

Woodley, S., Locke, H., Laffoley, D., MacKinnon, K., Sandwith, T., et al. (2019b) 'A review of evidence for area-based targets for the post-2020 global biodiversity framework'. *PARKS*, vol 25, no. 2, pp 31–46

7

HOW MUCH IS ALREADY SET ASIDE FOR CONSERVATION?

No overall figure for extent of area-based conservation exists. Data remain partial: best for protected areas, virtually non-existent or at best very rough estimates for other types of area-based conservation. Protected areas already cover 15 per cent of the land and 7.8 per cent of the ocean. Indigenous people have tenure rights on around a quarter of the world's land surface, but the amount of this that will remain as relatively natural ecosystem is still unknown; ICCAs are increasing, but there are no global statistics. OECMs are only just starting to be established, although there are already some studies of their likely extent. And regarding information on ecological representation, protected areas still have only very partial coverage for key biodiversity areas and at an ecoregional scale, many ecoregions remain under-represented or absent from the protected area estate.

We still have only a very partial picture of the total amount of area-based conservation in place around the world. As mentioned, no data (or at least only very generalised and approximate information) is available for many of the types discussed in this book. OECMs have at the time of writing only recently emerged as an agreed concept with an international definition and governments are still trying to work out what "counts"; it will take a number of years for this to settle down into anything like a global picture.

Areas under protection

Comparatively good information exists about protected areas thanks to the World Database of Protected Areas (WDPA, protectedplanet.net) compiled and run by the UNEP World Conservation Monitoring Centre (Bingham et al., 2019). Governments report a range of information about their protected area system to the WDPA, including name, size, date of creation, IUCN management category etc; this eventually feeds into the *UN List of Protected Areas*, the official global figures. Nonetheless, there are still many gaps and mistakes in the system. There is for example often a substantial time-lag between countries creating and reporting new protected areas, while occasional data clean-ups means that protected areas that have been counted twice (for instance if the name has changed) are removed and it looks as if there has been a net loss of area. Many governments only report state-run protected areas. Some countries also practice degazettement, an issue we will discuss later on.

The *Protected Planet Digital Report* (livereport.protectedplanet.net) is now updated monthly and is virtually a real-time information source. As of October 2019, it reports that the planet contains 241,368 protected areas. Just over 20 million km^2 are on land, equivalent to almost 15 per cent of the earth's land surface (excluding Antarctica) and just over 28 million km^2 or 7.8 per cent of the sea.

This means that an area of the world's terrestrial area and inland waters greater than South and Central America is now in protected areas (UNEP-WCMC et al., 2018). Well over half of these protected areas have been recognized since 1970 – a unique example of governments and other stakeholders consciously changing management approaches to land and water at a significant scale. Regions vary in the way that they designate protected areas: South America, Africa, Russia, Greenland and Australia tend to have some very large reserves (and may have proportionately less in number although this is not always the case), while other regions, and especially Europe, tend to have larger numbers of smaller reserves (UNEP-WCMC and IUCN, 2018). After a huge boom in rate of protection, there are signs that this is now slowing down (Gannon et al., 2019).

In the marine realm, after a slow start, things are currently moving more quickly, although global figures have been rather distorted by the designation of some very large marine protected areas. The Marae Moana (Cook Islands Marine Park) alone covers over 1.9 million km^2 for instance. However, as noted, some designations are controversial, with critics complaining that they have been set with so few restrictions that they exist in name only (De Santo, 2013). This may partly be the result of governments rushing to fulfil CBD agreements but also reflects huge battles with the fishing industry, with active resistance to anything that puts a curb on the catch in many parts of the world. Coastal areas are relatively better protected than the open ocean, where it is politically complicated to agree on marine protected areas; there is currently only 1.2 per cent protection of open oceans (UNEP-WCMC and IUCN, 2018).

There is clearly some under-reporting; as we write China has temporarily withdrawn most of its protected areas from the World Database on Protected Areas because it is re-designating and rethinking management approaches; a temporary change like this can distort global figures. Furthermore, some protected area types tend to be under-reported, including any managed by bodies other than the government. There are 878 privately protected areas listed on the WDPA from South America, whereas statistics from the region identify over 4,000 (Monteferri, 2019). On the other hand, as discussed, some governments report designations as "protected areas" that almost certainly do not meet the IUCN definition; statistics inevitably remain approximate as a measure of real conservation.

ICCAs are much less thoroughly studied, although there are some large country studies including a massive analysis undertaken in India (Pathak, 2009) and a similarly grand tour of marine ICCAs in the Philippines (Lavides et al., 2005). Both are now rather out of date. The ICCA Consortium estimates that global ICCA coverage (much of it not recognised officially as protected areas) will approximately equal in coverage the land area under state protected areas (Kothari et al., 2012). This figure is derived from calculations of territory under indigenous control, although there are some heroic assumptions. There is now an ICCA registry linked to the Protected Planet website, providing a secure, offline listing of sites (although detailed case studies are publicly available). But so far only a small number of ICCAs have registered.

At a more general level, research has found that indigenous people are stewards of or have tenure rights over around 38 million km^2 in eighty-seven countries; much or most of this land contains high biodiversity and 60 per cent is outside protected areas (Garnett et al., 2019). These statistics are fascinating and highlight the importance, which we return to in Part III, of recognising and working closely with many indigenous peoples, who hold territory containing the majority of the world's biodiversity. But such lands are also often referred to as if they are a done deal for biodiversity conservation. This is making a big assumption about how traditional custodians wish to use their lands: if they want to convert some or all of them to crops or sell them off to a ranching concerns or plantations, or to practice mining, it is their own business, and such changes do occur and will continue to occur. The romanticised view that all indigenous people will want to turn their backs on Western-style civilisation, smart phones, blue jeans and air conditioning is simplistic and insulting. As we have noted, these areas are also far from secure from changing government policy. So, while these areas are important and represent an opportunity for conservation and a more sustainable development, the future of many remains highly uncertain.

OECMs are still a largely theoretical designation, but given their official recognition in late 2018 this is likely to change very quickly. Some theoretical studies have already suggested what this might mean. Analysis of 740 terrestrial Key Biodiversity Areas in ten countries found that 76 per cent of those containing no protected areas were at least partly covered by potential OECMs (Donald et al., 2019).

As regards ecosystem services, no global figures exist so matching these to protected area remains impossible at the present time.

Are these areas in the best places?

Protected areas are of little practical benefit if they simply "protect" areas that are likely to be off limits for development anyway: the highest mountains, sand deserts, icefields and empty areas of the ocean. But these are also the areas that it is politically easiest to protect. The match between protected areas and sites believed to be important for biodiversity and/or ecosystem services is therefore less clear. For example, only 21 per cent of KBAs are completely covered by protected areas and

35 per cent contain no protected areas at all (UNEP-WCMC and IUCN, 2018). Whilst protected areas are not deemed essential for KBAs (IUCN, 2016), these statistics suggest that many KBAs remain vulnerable. In 2018, 43.2 per cent of the 821 terrestrial ecoregions (excluding Antarctica, rock and ice) met the CBD Aichi 11 target of at least 17 per cent of their area included within protected areas, but at the other extreme 5.6 per cent of ecoregions still had less than 1 per cent protected area coverage, or no protection at all (all figures, UNEP-WCMC and IUCN, 2018).

Similar disparities occur in the marine realm. Less than half the nearshore marine ecoregions have achieved the CBD's target of 10 per cent protection and only four out of 37 pelagic targets have reached 10 per cent protection (UNEP-WCMC and IUCN, 2018). Overlaying different marine prioritisation processes found that 88 per cent of the areas identified as important by four or more prioritisation initiative is not protected, with gaps occurring particularly in the Caribbean, Madagascar, Mediterranean, the Coral Triangle and the southern tip of Africa (Gownaris et al., 2019).

There is also a question of size. Large areas of natural ecosystem are far more resilient than smaller areas, which are likely to see biodiversity decline over time, unless they are well connected to other natural ecosystems. However, relatively small protected areas dominate the global protected area estate by number, although they remain relatively less well studied. Reviewing average size of almost 200,000 protected areas in the United States found that the average size is around 2,000 ha and the median around 20 ha (Baldwin and Fouch, 2018); we still don't know anything like enough about how these might be integrated into the conservation estate.

Do they form a coherent network?

Protected areas have in many cases been assembled piecemeal, as lands or waters became available driven by individual decisions and the earliest established at a time when little was understood about the dynamics of ecology. Connectivity between protected areas has apparently increased over the last decade (Saura et al., 2019), indicating clear improvements in the design of protected area systems in many countries. But as noted, a huge number of protected areas are at least partially isolated, cut off from other suitable ecosystems in ways that make it difficult or impossible for many species to cross. Even the largest and most mobile species face the risk of being cut off; the recently agreed "roadmap" for jaguar conservation in Latin America is based largely around the need to maintain ecological connectivity for the jaguar.

Conclusion: so how much is conserved?

At the moment, we don't know. Roughly 15 per cent of land and almost 8 per cent of ocean is in some form of officially recognised "protected area"; this represents a huge global commitment but as yet still does not cover a representative section of ecosystems and species, nor (probably to an even greater extent) does it secure all the ecosystem services that humanity requires. Additionally, if we take the estimates of Garnett and colleagues (2019) there are 23 million km² of indigenous peoples' territories outside protected areas, around 15 per cent of the land surface, an unknown proportion of which will remain in a reasonably natural condition in the long run. This is one of a number of estimates of indigenous territories; many of these areas are under threat. There are obviously also large areas of land (and water) that could qualify as OECMs, or play another conservation role, but we are still only just starting to assemble data; many of these will overlap with indigenous territories. In many parts of the world there are also large areas of degraded land, freshwaters and coastal waters that are of little use for productive purposes but might to some extent be restored. However, to repeat the statement at the start of this chapter, it is still fairly unclear what a global figure for area-based conservation might be.

References

Baldwin, R.F. and Fouch, N.T. (2018) 'Understanding the biodiversity contributions of small protected areas presents many challenges'. *Land*, vol. 8, no. 43, pp 87–98

Bingham, H.C., Juffe Bigoli, D., Lewis, E., MacSharry, B., Burgess, N.D., et al. (2019) 'Sixty years of tracking conservation progress using the world database on protected areas'. *Nature Ecology and Evolution*, vol. 3, pp 737–743

De Santo, E.M. (2013) 'Missing marine protected area (MPA) targets: How the push for quantity over quality undermines sustainability and social justice'. *Journal of Environmental Management*, vol. 124, pp 137–146

Donald, P.F., Buchanan, G.M., Balmford, A., Bingham, H., Couturier, A.R., et al. (2019) 'The prevalence, characteristics and effectiveness of Aichi target 11's "other effective area-based conservation measures" (OECMs) in key biodiversity areas'. *Conservation Letters*. Doi:10.1111/conl.12659

Gannon, P., Dubois, G., Dudley, N., Ervin, J., Ferroer, S., et al. (2019) 'Editorial essay: An update on progress towards Aichi biodiversity target 11'. *PARKS*, vol. 25, no. 2, pp 7–18

Garnett, S.T., Burgess, N.D., Fa, J.E., Fernández-Llamazares, A., Molnár, Z., et al. (2019) 'A spatial overview of the global importance of indigenous lands for conservation'. *Nature Sustainability*, vol. 1, pp 369–374

Gownaris, N.J., Santora, C.M., Davis, J.B. and Pikitch, E.K. (2019) 'Gaps in protection of important ocean areas: A spatial meta-analysis of ten global mapping initiatives'. *Frontiers in Marine Science*, vol. 6, no. 650. Doi:10.3389/fmars.2019.00650

IUCN. (2016) *A Global Standard for the Identification of Key Biodiversity Areas*, Version 1.0. IUCN, Gland, Switzerland

Kothari, A., Corrigan, C., Jonas, H., Neumann, A. and Shrumm, H. (eds.) (2012) *Recognising and Supporting Territories and Areas Conserved by Indigenous Peoples and Local Communities: Global Overview and National Case Studies*. Secretariat of the Convention on Biological Diversity, ICCA Consortium, Technical Series No. 64. Kalpavriksh and Natural Justice, Montreal, Canada

Lavides, M.N., Pajaro, M.G. and Nozawa, M.C. (eds.) (2005) *Atlas of Community-Based Marine Protected Areas in the Philippines*. Haribon Foundation, Quezon City, Philippines

Monteferri, B. (ed.) (2019) *Áreas de conservación privada en el Perú: avances y propuestas a 20 años de su creación.* Sociedad Peruana de Derecho Ambiental, Lima

Pathak, N. (ed.) (2009) *Community Conserved Areas in India: A Directory.* Kalpavriksh, Pune, India

Saura, S., Bertzky, B., Bastin, L., Battistella, L., Mandrici, A., et al. (2019) 'Global trends in protected area connectivity from 2010 to 2018'. *Biological Conservation*, vol. 238. Doi:10.1016/j.biocon.2019.07.028

UNEP-WCMC and IUCN. (2018) *2018 United Nations List of Protected Areas: Supplement on Protected Area Management Effectiveness.* UNEP World Conservation Monitoring Centre, Cambridge

UNEP-WCMC, IUCN and NGS. (2018) *Protected Planet Report 2018.* UNEP-WCMC and IUCN, Cambridge and Gland, Switzerland

PART III
Why this is not enough

8

THE NEEDS FOR AREA-BASED CONSERVATION

Countering threats

Most people reading this book will know about the threats facing the environment and so these are dealt with only in summary. Biodiversity is declining at unprecedented rates; most recent estimates place a million species at serious risk. Other ecosystem services are also declining, leading in turn to loss of food and water security. Furthermore, public and political awareness of the role that these services play is often so low that problems are ignored until it is too late. Population increases, rising consumption and mounting levels of waste are creating natural resource shortages. High levels of land degradation, natural ecosystem loss and decline in freshwaters are being recorded; climate change is exacerbating many pressures and leading to some frightening new problems, including ocean acidification. Poaching has reached crisis levels in some countries. The concept of planetary boundaries is becoming better known, with scientists estimating the point at which use of particular resources, emission of pollutants or other ecological disruptors reach an unsustainable level and "exceed the boundary". All of these issues have direct implications for human security, both today and in the future.

As we learn more about the world around us, and as society changes, our reasons for setting aside areas of land and water in a more-or-less natural state are changing as well (Watson et al., 2014). Although conservation of wildlife and biodiversity has dominated discussions about protected area establishment for the last few decades, it may not be the most significant factor in terms of the total area that society might choose to set aside. Other ecosystem services, including those relating to reducing and adapting to climate change could well turn out to be more important, at least for the next fifty years or more, although like other factors this could also change over time. Chapter 9 looks in more detail at the suite of benefits that protected areas provide. In the following, we first summarise some of the problems that have created the need for protected areas. This will be fairly brief; there are numerous, miserable, dystopian accounts of our likely future problems, and our focus here is on trying to address some of these. But nevertheless, it is important to understand the wider reasons why we might wish to set aside large areas of natural ecosystems.

Converging crises: biodiversity, ecosystem services, climate and approaching planetary boundaries

As conservation professionals we must always be careful about crying "wolf" too often; the public grows wary of threats of doom and disaster, and not every disaster comes to pass. Reading back through some of the older volumes on our bookshelves it seems as if we should all have starved to death years ago, run out of oil or been literally crowded off the planet.

But bearing that caution in mind, the evidence is mounting of a genuine crisis underway, one that is having and will continue to have direct consequences for human society as well as impacting heavily on the rest of nature. The recent assessment from the Intergovernmental Science-Policy Platform on Biodiversity and Ecosystem Services (IPBES, 2019b) is only the latest to point to a continuing catastrophic rate of biodiversity loss, with a million species estimated to be at risk. Most of these will still never have been recognised or described by scientists; a myriad of small plants and invertebrates that help make up the astonishing richness of the planet's ecosystems. But many larger and more iconic species are at risk as well, their decline gloomily captured in successive assessments of the IUCN Red List. Tiger populations are little more than 4,000 individuals in the wild (Goodrich et al., 2015), an astonishing 97 per cent decline in a century (Walston et al., 2010). Half the world's lion population has disappeared in the last twenty years, and viable populations are now confined mainly to ten specific regions of sub-Saharan Africa (Bauer et al., 2015). Ten bat species affected by white nose syndrome have suffered up to 90 per cent decline in hibernating individuals in parts of North America (Bure and Moore, 2019), and extinctions amongst frogs and other amphibians are running at shocking levels (Stuart et al., 2008); in both cases novel diseases have wiped out huge swathes of individuals. The problems are not confined to animals; plants are becoming extinct in large numbers as well, with almost 600 recorded extinctions believed to be a small percentage of the total (Humphreys et al., 2019). According to an analysis by the Kew Botanic Gardens in London, one in five plant species is threatened with extinction (2016).

And these declines are in most cases from an already severely impoverished base. Prehistoric extinctions of megafauna by our human and close-to-human ancestors has already destabilised ecosystems; although these changes seem way in the past to us, in evolutionary terms they happened a few minutes ago with no time for an effective rebalancing. More recently, massive levels of hunting and fishing have depressed populations so that current losses are from populations already a fraction of the original. Nineteenth-century fishing communities have records of abundance that today's fleets can only dream about (Roberts, 2012). Around our village in Wales, records from 150 years ago talk of "bags" of dozens of hares killed in a single day; nowadays if we see two or three in a year, we are excited. When the number of individuals of a particular species declines too far it can enter a downward spiral towards extinction that is very hard to reverse without expensive, time-consuming interventions and a good deal of luck.

We're not going to labour the point; most people reading this will be aware of both the scale and seriousness of the situation. More shocking still is recent evidence that impacts are spreading even into areas that were once considered safe, with for instance dramatic declines of insects found in nature reserves in Germany (Hallmann et al., 2017) and protected forests in Puerto Rico (Lister and Garcia, 2018). Meta-studies show up to 40 per cent of insect species threatened with extinction in the next few decades (Sánchez-Bayo and Wyckhus, 2019). Such evidence raises serious questions about a strategy of intensifying land use in one area and setting aside other areas for conservation; many impacts do not respect artificial borders set up by land-use planners. We will look at these issues in more detail when considering the effectiveness of area-based conservation.

Threats to ecosystem services

It is not just species that are risk; many wider environmental pressures directly threaten food and water security, increase risks of weather-related disasters and disrupt societies. In 2017 the UN Convention to Combat Desertification (UNCCD) produced the *Global Land Outlook*, the first attempt by a UN agency to look at the status of land overall (UNCCD, 2017). A year later, the Ramsar Convention on Wetlands produced an equivalent *Global Wetlands Outlook* (Ramsar, 2018), considering the status of inland waters, peatlands and coastal regions. We were heavily involved in both efforts, providing an up to date picture of what we know about the status of the planet outside of the open oceans. The evidence is sobering – evidence not only of the scale of loss but also of the way in which many people have not even fully observed or recognised that a loss has taken place.

The pressures on global land resources are larger than at any other time in human history. A rapidly increasing human population is placing mounting demands on the land base and on freshwater and marine resources. The situation is exacerbated by rising consumption levels in many parts of the world; every child born in the rich countries of the west will consume the same as many children in the poorer countries. Expectations are changing, particularly regarding the types of food that people want to eat and the proportion of meat in the diet. These factors result in growing competition for land use, especially between the desire for a range of different foods, fibres and fuels and the maintenance of basic life support systems on the planet.

In the starkest terms, there is direct competition between the demand for provisioning services that benefit people, like food, water and energy, and the need to protect other ecosystem services that support all life on Earth. This means that overall ecosystem services are either being used at unsustainable rates or are being actively undermined as a result of human activities (IPBES, 2019b). The chairman of the IPBES, Sir Robert Watson, put it this way: "We are eroding the very foundations of our economies, livelihoods, food security, health and quality of life worldwide" (IPBES, 2019a). Biodiversity underpins all these services (Mace et al., 2011) and underwrites the full enjoyment of a wide range of human rights, such as the rights to a healthy life, nutritious food, clean water and cultural identity.

A significant proportion of both production lands and natural ecosystems is degrading in quality. From 1998 to 2013, approximately 20 per cent of the Earth's vegetated land surface showed persistent declining trends in productivity, evident in 20 per cent of cropland, 16 per cent of forest land, 19 per cent of grassland and 27 per cent of rangeland. These trends are especially alarming in the face of increased demand for land-based goods and services and are being worsened by climate change (UNCCD, 2017). Rather than manage existing crop and grazing land sustainably we are using it up and moving on to clear more areas of natural vegetation, leaving abandoned or sub-optimal land behind that will in time require expensive restoration efforts.

Land degradation also triggers competition for scarce resources, leading to migration and insecurity while exacerbating access and income inequalities. Soil erosion, desertification and water scarcity all contribute to societal stress and breakdown. Here land degradation can be considered a "threat amplifier" (van Schaik and Dinnissen, 2014), especially when it slowly reduces people's ability to use land for food production and water storage or undermines the wider ecosystem services supplied by a healthy land base. This in turn may trigger or increase the risk of conflict.

Land degradation itself further contributes to climate change by releasing greenhouse gases into the atmosphere. Current management practices in the land-use sector are responsible for a quarter of the world's greenhouses gases, a total that has remained constant for some time (Smith et al., 2014). Land degradation also increases the vulnerability of people, especially the poor, women and children. Over 1.3 billion people, mostly in the developing countries, are trapped on degrading agricultural land exposed to climate stress and therefore largely unable to benefit from wider economic growth. A new reality of "rural transformation" is replacing the long-held expectation of rural development as people abandon the land and migrate to urban areas – an extraordinary phenomenon that has changed and often impoverished cultural identity and permanently altered landscapes, yet one that has gone largely unremarked (UNCCD, 2017). Furthermore, it may well be that these changes are only really starting to accelerate. For example, worldwide an estimated 410 million farms are less than a hectare in size; their future remains deeply uncertain (Dudley and Alexander, 2017), with all the trends suggesting consolidation of farmland into huge units owned by people based many miles away, often in different countries.

Degradation is one thing and often reversible, given time and money. Complete conversion is another matter, and much more difficult to undo. While deforestation is slowing in some areas, it is accelerating in others (Taylor, 2015). Despite all the publicity and attention paid to forest loss, deforestation has actually been accelerating in the last few years throughout the tropics. Net rate of loss averaged 3.3 million hectares a year between 2010 and 2015, with 12 million hectares being lost in 2019 alone (Weisse and Dow Goldman, 2019). Rates of destruction of natural forests are even higher than many global statistics suggest, as native forests are displaced by monoculture plantations both of which are classified as "forests" in official statistics although their ecosystem services are very different (FAO, 2016). And rates of losses of grassland and savannahs generally attract much less attention than deforestation but may even be happening more quickly as areas like the Cerrado in Brazil and the pastures of Argentina, Uruguay and Paraguay are ploughed up to plant soy and other crops to feed livestock.

Wetlands are also changing rapidly, with many areas drained, river courses altered particularly through dam construction, while pollution and overfishing are impoverishing aquatic life. Global inland and coastal wetlands cover over 12.1 million km², with 54 per cent permanently inundated and 46 per cent seasonally inundated. Natural wetlands are in long-term decline; between 1970 and 2015, inland and marine/coastal wetlands both declined by approximately 35 per cent, where data are available, three times the rate of forest loss. In contrast, human-made wetlands, largely rice paddy and reservoirs, almost doubled over this period, now forming 15 per cent of wetlands. These increases have not compensated for natural wetland loss (Ramsar Convention, 2018).

The world's oceans are arguably under greater threat than any terrestrial or freshwater ecosystem, doubly hampered by being out of sight and often therefore out of mind, threatened by chronic overfishing, pollution and ocean acidification as a result of climate change. Unsustainable fishery has been identified as the single largest threat to marine ecosystems (Pauly et al., 2005) although ocean acidification may since have become even more significant and potentially catastrophic (Laffoley and Baxter, 2016). Many fish farms also rely on wild-caught fish for feed (Naylor et al., 2000), increasing overall pressure on natural populations. Up to a third of corals face increased extinction threat under climate change (Carpenter et al., 2008) with threats of ocean acidification further increasing pressure on corals (Hoegh-Guldberg, 2007) and many other types of marine life.

Climate change is so well known we will discuss it only briefly here, except to say that the majority of evidence at present suggests that of the many scenarios produced to project likely future conditions, the ones predicting the more severe impacts seem to be emerging as most likely; conditions and ecosystems are changing faster and more substantially than scientists have often predicted (Lorenz et al., 2019), and there are increasing fears of feedback loops further increasing the rate and severity of climate change above predicted levels (Steffen et al., 2018). Ice melt appears to be occurring faster in the Arctic (Serreze and Stroeve, 2015), sea-level rise is accelerating above many of the past projections (Chen et al., 2017), most of the world's glaciers are shrinking fast (Milner et al., 2017), and the world's weather is becoming both more extreme and far less predictable. We first investigated the likely impacts of climate change on biodiversity in 1993 (Markham et al., 1993), assembling the predictions of ecologists working on these issues from around the world. In the quarter century since, nothing has reduced our pessimism about the scale of likely impacts. If anything, we were too cautious; for instance, the devastating effects of increased fire frequency and hotter fires under climate change already being seen in many countries were largely unpredicted.

Other forms of pollution are also important. Acid deposition caused major impacts in parts of Europe, North America and Asia in the mid-late twentieth century, particularly to lichens, mosses, freshwater fish and other aquatic species but also to a range of other invertebrates and plants (Dudley et al., 1984). While increased pollution control and a reduction in coal burning has generally reduced these impacts, freshwater acidification has been increasing in places such as the North American Great Lakes (Phillips et al., 2015). Pesticide drift can be a problem when agricultural areas are found next to reserves and catastrophic levels of invertebrate decline are being seen in parts of Europe (Vogel, 2017). Mining, particularly illegally mining, can lead to dramatic and dangerous levels of mercury and arsenic, including inside protected areas, for example in the Amazon (Ouboter et al., 2012). There is also a growing awareness of plastic pollution, which is not only rapidly increasing but cannot be mitigated by protection due to its ability to travel long distances from source, particularly in water bodies. Even the most remote marine protected areas are seeing rapidly increasing levels of plastics and evidence of their impacts, a major threat to our marine environment (Barnes et al., 2018).

Invasive species are also a major threat to biodiversity, changing ecology, eliminating vulnerable species and generally reducing overall biodiversity (Foxcroft et al., 2013). A non-native species only becomes invasive under certain circumstances, which boil down to lack of natural control mechanisms allowing rapid population growth and spread. In the case of invasive plants, this can mean that native species get swamped and pushed aside and wider ecological changes can result. The role of water hyacinth in blocking waterways throughout Asia and Africa, following its spread from its native Latin America, is a classic example (Villamagna and Murphy, 2010). In the case of invasive animals the chief problems come from predation, especially when native species do not have adequate defence mechanisms; weasels and stoats in New Zealand, mink in the UK and snakes on the Pacific Island of Guam have all led to the extinction or near-extinction of native species.

Dwindling resources

It is not only species and ecosystems that are under pressure; many other resources are also reaching critical levels in a parallel and frequently connected process of decline. Johan Rockström and colleagues have identified a set of "planetary boundaries" that they believe should not be exceeded if a sustainable future is to be assured, covering issues like climate change, biodiversity loss, nitrogen and phosphorus use and others (e.g. Rockström, 2009); several of these proposed boundaries are already being exceeded.

While politicians continue to squabble and equivocate, and polluting industry invests millions in misinformation, at day-to-day-level farmers, forest managers and billions of the world's most vulnerable people are trying to make do in a world that no longer seems to be secure. It is slightly unreal to contrast two worlds here. People we meet in the field – poor farmers, fisherfolk and those scraping by on the edge of existence as well as conservation professionals – are generally now all stoically making plans to adapt to the changing conditions they see around them. On the other hand, we listen to the arguments of politicians and political parties, staunchly in denial that climate change is a problem or equivocating about taking any steps to address the problem that might inconvenience them in the slightest.

All of the factors described earlier combine to reduce the ecosystem services that we rely on, such as contributions to food and water security, disaster risk reduction, health benefits and a host of hard-to-measure aesthetic, cultural and spiritual qualities. Responses need to be wide-ranging and, in some cases, radical and need to be enacted at both the highest levels of politics and business and in everyone's day-to-day lives. In the following chapters we review the role that an expanded area-based conservation programme can play.

References

Barnes, D.K.A., Morley, S.A., Bell, J., Brewin, P., Brigden, K., et al. (2018) 'Marine plastics threaten giant Antarctic marine protected areas'. *Current Biology*, vol. 28, pp 1121–1142. Doi:10.1016/j.cub.2018.08.064

Bauer, H., Chapron, G., Nowell, K., Henschel, P., Funston, P., et al. (2015) 'Lion (*Panthera leo*) populations are declining rapidly across Africa, except in intensively managed areas'. *Proceedings of the National Academy of Sciences*, vol. 112, pp 14894–14899

Bure, C.M. and Moore, M.S. (2019) 'White-nose syndrome: A fungal disease of North American hibernating bats'. In: White, W.B., Culver, D.C. and Pipan, T. (eds.) *Encyclopedia of Caves*, Elsevier, Chapter 36, pp 1165–1179

Carpenter, K.E., Abrar, M., Aeby, G., Aronson, R.B., Banks, S., et al. (2008) 'One-third of reef-building corals face elevated extinction risk from climate change and local impacts'. *Science*, vol. 321, pp 560–563

Chen, X., Zhang, X., Church, J.A., Watson, C.S., King, M.A., et al. (2017) 'The increasing rate of global mean sea-level rise during 1993–2014'. *Nature Climate Change*. Doi:10.1038/NCLIMATE3325

Dudley, N. and Alexander, S. (2017) 'Will small farmers survive the 21st century – and should they?' *Biodiversity*. Doi:10.1080/14888386.2017.1351397

Dudley, N., Barrett, M. and Baldock, D. (1984) *Th Acid Rain Controversy*. Earth Resources Research, London

FAO. (2016) *Global Forest Resource Assessment (GFRA) Summary 2015*. Food and Agriculture Organisation, Rome

Foxcroft, L.C., Pyšek, P., Richardson, D.M. and Genovesi, P. (eds.) (2013) *Plant Invasions in Protected Areas: Patterns, Problems and Challenges*. Springer, Dordrecht

Goodrich, J., Lynam, A., Miquelle, D., Wibisono, H., Kawanishi, K., et al. (2015) 'Panthera Tigris'. *The IUCN Red List of Threatened Species*. http://dx.doi.org/10.2305/IUCN.UK.2015-2.RLTS.T15955A50659951.en

Hallmann, C.A., Sorg, M., Jongejans, E., Siepel, H., Hofland, N., et al. (2017) 'More than 75 percent decline over 27 years in total flying insect biomass in protected areas'. *PLoS One*, vol. 12. https://doi.org/10.1371/journal. pone.0185809

Hoegh-Guldberg, O., Mumby, P.J., Hooten, A.J., Steneck, R.S., Greenfield, P., et al. (2007) 'Coral reefs under rapid climate change and ocean acidification'. *Science*, vol. 318, pp 1737–1742

Humphreys, A.M., Govaerts, R., Ficinski, S.Z., Lughadha, E.N. and Vorontsova, M.S. (2019) 'Global dataset shows geography and life form predict modern plant extinction and rediscovery'. *Nature Ecology and Evolution*, vol. 3, pp 1043–1047. Doi:10.1038/s41559-019-0906-2

IPBES. (2019a) *Nature's Dangerous Decline "Unprecedented", Species Extinction "Accelerating"*. Media Press Release, Intergovernmental Science-Policy Platform on Biodiversity and Ecosystem Services, 7 May

IPBES. (2019b) *Summary for Policymakers of the Global Assessment Report on Biodiversity and Ecosystem Services of the Intergovernmental Science-Policy Platform on Biodiversity and Ecosystem Services*. IPBES Secretariat, Bonn, Germany

Kew. (2016) *State of the World's Plants 2016*. Royal Botanic Gardens Kew, London

Laffoley, D. and Baxter, J (eds.) (2016) *Explaining Ocean Warming: Causes, scale, effects and consequences*. IUCN, Gland, Switzerland

Lister, B.C. and Garcia, A. (2018) 'Climate-driven declines in arthropod abundance restructure a rainforest food web'. *Proceedings of the National Academy of Sciences*, vol. 115, pp e10397–e10406

Lorenz, R., Stalhandske, Z. and Fischer, E.M. (2019) 'Detection of a climate change signal in extreme heat, heat stress, and cold in Europe from observations'. *Geophysical Research Letters*. Doi:10.1029/2019GL082062

Mace, G.M., Norris, K. and Fitter, A.H. (2011) 'Biodiversity and ecosystem services: A multilayered relationship'. *Trends in Ecology and Evolution*, vol. 27, pp 19–26

Markham, A., Dudley, N. and Stolton, S. (1993) *Some Like It Hot: Climate Change, Biodiversity and the Survival of Species*. WWF International, Gland, Switzerland

Milner, A.M., Khamis, K., Battin, T.J., Brittain, J.E., Barrand, N.E., et al. (2017) 'Glacier shrinkage driving global changes in downstream systems'. *Proceedings of the National Academy of Sciences*, vol. 114, pp 9770–9778. Doi:10.1073/pnas.1619807114

Naylor, R.L., Goldburg, R.J., Primavera, J.H., Kautsky, N., Beveridge, M.C.M., et al. (2000) 'Effect of aquaculture on world fish supplies'. *Nature*, vol. 405, pp 1017–1024

Ouboter, P.E., Landburg, G.A., Quik, J.H.M., Mol, J.H.A. and van der Lugt, F. (2012) 'Mercury levels in pristine and gold mining impacted aquatic systems in Suriname, South America'. *Ambio*, vol. 41, pp 873–882

Pauly, D., Watson, R. and Alder, J. (2005) 'Global trends in world fisheries: Impacts on marine ecosystems and food security'. *Philosophical Transactions of the Royal Society B*, vol. 360, pp 5–12

Phillips, J.C., McKinley, G.A., Bennington, V., Bootsma, H.A., Pilcher, D.J., et al. (2015) 'The potential for CO_2-induced acidification in freshwater: A great lakes case study'. *Oceanography*, vol. 28, no. 2, pp 136–145

Ramsar Convention. (2018) *Global Wetland Outlook*. Ramsar Convention, Gland, Switzerland

Roberts, C. (2012) *Ocean of Life: How Our Seas Are Changing*. Penguin, London

Rockström, J. (2009) 'A safe operating space for humanity'. *Nature*, vol. 461, pp 472–475

Sánchez-Bayo, F. and Wyckhus, K.A.G. (2019) 'Worldwide decline of the entofauna: A review of the drivers'. *Biological Conservation*, vol. 232, pp 8–27

Serreze, M.C. and Stroeve, J. (2015) 'Arctic sea ice trends, variability and implications for seasonal ice forecasting'. *Philosophical Transactions of the Royal Society A*, vol. 373. Doi:10.1098/rsta.2014.0159

Smith, P., Bustamante, M., Ahammad, H., Clark, H., Dong, H., et al. (2014) 'Agriculture, forestry and other land use (AFOLU)'. In: Edenhofer, O., Pichs-Madruga, R., Sokona, Y., Farahani, E., Kadner, S., et al. (eds.) *Climate Change 2014: Mitigation of Climate Change.*

Contribution of Working Group III to the Fifth Assessment Report of the Intergovernmental Panel on Climate Change. Cambridge University Press, Cambridge and New York

Steffen, W., Rockström, J., Richardson, K., Lenton, T.M., Folke, C., et al. (2018) 'Trajectories of the earth system in the Anthropocene'. *Proceedings of the National Academy of Sciences*, vol. 115, pp 8252–8259

Stuart, S.N., Hoffmann, M., Chanson, J.S., Cox, N.A., Berridge, R.J., et al. (2008) *Threatened Amphibians of the World*. Lynx Edicions, IUCN and Conservation International, Arlington, VA

Taylor, R. (ed.) (2015) 'Saving forests at risk'. In: *WWF Living Forests Report*. WWF, Gland, Switzerland

UNCCD. (2017) *Global Land Outlook*. UN Convention to Combat Desertification, Bonn, Germany

van Schaik, L. and Dinnissen, R. (2014) *Terra Incognita: Land Degradation as Underestimated Threat Amplifier*. Netherlands Institute of International Relations. Clingendael, The Hague

Villamagna, A.M. and Murphy, B.R. (2010) 'Ecological and socio-economic impacts of invasive water hyacinth (*Eichhornia crassipes*): A review'. *Freshwater Biology*, vol. 55, pp 282–298

Vogel, G. (2017) 'Where have all the insects gone?' *Science*, vol. 356, no. 6338, pp 575–579

Walston, J., Robinson, J.G., Bennett, E.L., Breitenmoser, U., da Fonseca, G.A.B., et al. (2010) 'Bringing the tiger back from the brink – the six percent solution'. *PLoS Biology*, vol. 8. Doi:10.1371/journal.pbio.1000485

Watson, J.E.M., Dudley, N., Hockings, M. and Segan, D. (2014) 'The performance and potential of protected areas'. *Nature*, vol. 515, pp 67–73

Weisse, M. and Dow Goldman, E. (2019) *The World Lost a Belgium-Sized Area of Primary Rainforests Last Year*. World Resources Institute, 29 April. www.wri.org/blog/2019/04/world-lost-belgium-sized-area-primary-rainforests-last-year. Accessed 31 May 2019

9

THE NEEDS FOR AREA-BASED CONSERVATION

Maximising ecosystem services

Natural and semi-natural ecosystems do not just protect biodiversity; they also maintain a wide range of other ecosystem services — benefits that humans get from nature — critical for sustainable development. Many are provisioning services. Water security is helped by the better quality and sometimes increased quantity of water flowing from protected forests and wetlands. Food security is strengthened because protected areas maintain wild populations of food species, particularly fish, and support pollination, soil formation and crop and livestock wild relatives. Protected areas supply natural medicines, pharmaceutical products and building materials like rattan and bamboo. Regulatory services are also important; carbon storage and sequestration, and disaster risk reduction by buffering against extreme weather events and earth movements. Cultural services range from the historical and spiritual values of ecosystems or landscape features to the huge economic benefits of tourism, now close to the world's largest "industry". Many benefits have economic returns, but others do not, and ecosystem services cannot be judged in economic terms alone. Finally, many so-called natural ecosystems are the traditional homes of indigenous people or other local communities, who have practised lifestyles so close to nature that they have little impact on ecology. Many of these societies are now under severe threat; protected areas can also provide these people with a sanctuary. Scientists are starting to calculate exactly what ecosystems services supply, what we have lost and what we need to regain; this will be an increasingly important aspect of protected area management in the future.

Missing ecosystem services

A river runs through the valley we live in, heading down to the estuary and the Irish Sea. It rises in the Aran Mountains a few miles away and only measures about forty miles from source to sea, much of it within the Dyfi Valley biosphere reserve. Further downstream from our village there is a series of strict nature reserves protecting water birds and other wildlife. Immediately opposite our village, the fields flood regularly, when a high tide backs up the river and heavy rain has been falling in the hills, a far from unusual scenario in Wales. The farmers are adept at moving sheep off the grazing fields just before the river bursts its banks. The Dyfi Valley therefore acts as an "ecosystem service", the floodplain providing a harmless place to disperse excess water and saving communities downstream from flooding and other damage; an economist could fairly easily work out the monetary value of avoided damage. (A couple of valleys away, on the outskirts of the university town of Aberystwyth, the Welsh government built a regional headquarters on the floodplain and are now spending huge amounts of money trying to sort the problem out.) But there are only a handful of people in the valley who would even recognise the term "ecosystem service", let alone know that they were looking out at one every day. This isn't their fault; these issues still remain the purview of specialists. Ecosystem services are often only really recognised once they are lost, and not always even then. **ND/SS**

One of the challenges in advocating for area-based conservation is that it is all-too-often presented as a Manichean choice between nature and development. Asking a poor farmer or a landless labourer to ignore vast areas of prime forest land or savannah in favour of "wildlife", or even worse an obscure term like "biodiversity", is inevitably a hard sell. Even in the United States, at the time the richest country in the world, the protection of old-growth forest in the Pacific Northwest in the 1990s was commonly presented as a straight conflict between owls and logging families. Or more precisely, the survival of the northern spotted owl (*Strix occidentalis caurina*), an obscure sub-species of a bird very few people will ever see, and the economic livelihood of logging communities in parts of the country where few other employment alternatives exist (Caufield, 1990). In fact, biodiversity conservation is one of a long list of benefits that we derive from natural and semi-natural ecosystems, some of which probably require more space than would be needed for protecting biodiversity alone (Larsen et al., 2014). It often suits the purposes of those promoting land use change to argue that conservationists are just harking back to a romanticised past or putting neo-colonialist attitudes to wildlife above the needs of local people. It's not as simple as that.

We started to look at the wider benefits of protected areas around the turn of the century in what we first thought would be a fairly quick project. The research stretched into something much bigger and we're still working on aspects of this today. Our initial study resulted in a series of reports, *Arguments for Protection*, in collaboration with WWF, the World Bank and other partners and eventually a book (Stolton and Dudley, 2010). But global understanding of and experience with managing ecosystem services is developing very quickly, and in the following chapter we will attempt both to summarise and update what we have learned.

Like many of the terms used in this book "ecosystem service" has its detractors. The Millennium Ecosystem Assessment (MEA) first popularised it twenty years ago; more recently IPBES has coined the term "Natures' Contributions to People", while IUCN prefers Nature-Based Solutions and we and others use "natural solutions" to refer to those ecosystem services arising from area-based conservation. This covers a very wide field and the MEA divides these benefits into four. "Supporting services" are those fundamental to maintaining life on the planet: photosynthesis capturing solar energy, soil formation, nutrient cycling and the global water cycle. Without all these things we would be lost. Next is a large group of "provisioning services", including anything we extract from healthy natural and semi-natural ecosystems such as food, water, medicines, building materials and so on. Provisioning ecosystem services are things renewable if managed well; so for instance wood is renewable but extraction of minerals like fossil fuels is not, because in human terms it is a finite resource even if originally derived from living things. The third group is "regulating services", like water and air purification by natural processes, carbon sequestration and storage, various forms of natural buffering against extreme weather events, and pollination by insects and birds. Lastly is a complex group of "cultural services", which describes mainly non-extractive uses like recreation and aesthetic appreciation and include, critically, spiritual values and the importance of particular places and species to various faith groups.

IPBES has been critical of the MEA typology because it implies hard divisions, whereas in reality many benefits could fit partly into two or more of the sub-sets. This is true, but we have found the MEA typology a useful summary to help people realise the wide scope of ecosystem services, all the way from mitigating climate change to providing spiritual or emotional comfort. But IPBES is correct that many services do not fit neatly into any single category. For simplicity's sake here we divide them instead into the main benefits derived, which from the perspective of protected areas are water security, food security, health benefits, disaster risk reduction, contribution to climate change strategies, recreation and tourism, and a suite of cultural, aesthetic and spiritual values. Many include elements of many or all of the MEA divisions. Lastly, the economic implications of ecosystem services from protected areas will be briefly reviewed.

But before that, we'll have a brief look at some of the services that underpin all the others but are all too frequently ignored. "Supporting services" are so obvious that they are often taken for granted; without photosynthesis we would have no life for instance. But it is worth noting that sober and normally cautious scientists are increasingly questioning whether some of these services are now reaching a tipping point. Soil is a constantly renewing resource, made up of decaying organic matter and minerals from eroding rocks. Erosion is to be expected, with new soil replacing that blown or washed into the world's rivers and oceans. But it is now widely recognised that this whole system is out of balance, with erosion often exceeding formation leading to land degradation or even desertification. Furthermore, salinisation through poor irrigation practice is making huge areas of land unworkable, and organic matter levels are collapsing wherever agriculture relies on agrochemicals to the exclusion of recycling organic matter. Global figures are hard to calculate, but the UN Convention to Combat Desertification estimates that around a fifth of the world's agricultural lands have shown signs of persistent declining trends in productivity over the past twenty years (UNCCD, 2017). Changes at a similar scale are affecting many global hydrological systems while ocean acidification as a result of climate change is creating massive changes with impacts that are hard – and terrifying – to predict. Supporting services can no longer be taken for granted, but their loss is incalculable.

Protected areas will not solve these problems on their own. But retaining large areas of natural ecosystem helps keep things in some sort of balance. For example, at the time of writing the rate of Amazon clearing has just rocketed up to catastrophic levels. If climatologists are correct, major losses to the Amazon rainforest would have huge impacts on agricultural productivity farther south in the continent; populism and an anti-conservation agenda in Brazil could wipe out agricultural profits over huge areas of Argentina and Uruguay (Nobre, 2014). The role of natural ecosystems in maintaining supporting services deserves more attention than it gets.

Water security

Back in 2003, we coordinated a report on the role that forest protected areas play in providing drinking water and found that over a third of the world's hundred largest cities drew a significant proportion of their drinking water from protected areas. Water authorities in some, like Melbourne and New York, are completely aware of this fact and work actively with protected area agencies in consequence; indeed New York famously protected areas of the Catskill Mountains explicitly to maintain pure water supplies (Dudley and Stolton, 2003). This statistic has been endlessly recycled and joined the mass of other evidence for the role of natural ecosystems in providing drinking water, particularly the widespread recognition of forested mountains as "water towers" for downstream communities (e.g., Bhandari et al., 2008). But have changing perceptions really led to substantial policy changes?

Watersheds containing natural ecosystems, particularly forests and freshwaters, usually produce cleaner water than those flowing through agricultural land, intensive pasture or industrial and urban areas. This is partly for the obvious reason that natural systems have less pollution – nitrate fertilisers, pesticides and industrial wastes – and partly because of natural detoxification in these ecosystems. Then on top of that *some* ecosystems increase the net water flow, particularly tropical cloud forests, the páramos vegetation of the middle Andes and perhaps some old-growth Eucalyptus forests. They do this because in cloudy or misty conditions water condenses onto leaves and runs into the catchment; other forests have the reverse effect by transpiring and sending more water into the atmosphere (Hamilton et al., 2008). Third, forests, wetlands and rich native grasslands all absorb water and help smooth out the highs and lows caused by drought and downpour; providing more water during dry periods and reducing flooding during wet periods. Any natural ecosystem provides those services but protected areas help ensure that natural ecosystems remain intact, or intact enough to supply the services.

Water security from a protected area in Colombia

Chingaza National Park in Colombia is a case in point. Situated a short drive away from the capital, Bogotá, the protected area covers 766 km² of the eastern Andes, rising to over 4,000 metres, almost entirely within the Orinoco catchment. The protected area is part of the páramo ecosystem, unique to the Andes, dominated by many species of frailejones (*Espeletia* spp.), tall, slow-growing spikes of plants with hairy, succulent leaves that give the landscape a unique aspect. Frailejones are in many places under threat from clearance, particularly to plant potatoes, emphasising the importance of the national park. There are also around forty natural glacial lakes, some of which are sacred to the Muisca people. (Indeed, lakes in the páramos retain some sacred values to local people even today and I've come across groups clearly visiting for spiritual purposes in other parts of the region.) The park is significant for the species it contains; spectacled bear (*Tremarctos ornatus*) are found there along with tapirs (*Tapirus terrestris*), puma (*Puma concolor*), ocelots (*Leopardus pardalis*) and the Andean condor (*Vultur gryphus*), plus a mass of plants, many of which are still not fully surveyed. But Chingaza also plays a key role in maintaining the region's water security. This part of the Andes is perennially wet and in particular misty; when I was there we could only see a short distance due to the density of mountain cloud. Plants like frailejones are designed to scavenge water on their fleshy leaves and a glance at the top in anything but the driest conditions will find water sitting in tiny pools at the base of the leaves. Many species of peat moss also absorb huge amounts of water. Eventually this "captured" water seeps down into waterways, making Chingaza the supplier of 80 per cent of Bogotá's water. The municipal water company recognises this and pays for some of the protected area management costs, as does a soft drink bottling company situated at the base (although in the latter case only a fairly derisory sum). But despite the best efforts of Parques Nacionales Naturales, the state protected area agency, to explain the links, very few citizens of the capital realise that they are relying on a national park for their drinking water. **ND**

What has changed since we first looked at this around the turn of the century? Unfortunately, some of the benefits we identified then have been reduced; the two national parks that provide drinking water to Jakarta have been degraded with detrimental impacts on water supply for instance. Several of the water towers in Africa have undergone serious deforestation. Some things haven't progressed. The forest that has supplied drinking water for Istanbul since the early days of the city that was then named Constantinople was unprotected and at risk then, and it still is today. But pleasingly, most of the water towers identified then are still doing their job.

An undeniable change over the last twenty years is the colossal increase in both the number and size of cities. For example, Africa is experiencing the highest rate of urbanisation in the world. It already has over a third of its billion inhabitants living in urban areas, and this urban population is expected to triple by 2050. In 1960, there were only five cities in sub-Saharan Africa with over half a million inhabitants, by 2015 there were eighty-four, and by 2030 there will likely be 140 (Satterthwaite, 2014). Some of the continent's largest cities draw water from protected areas: Nairobi from Aberdares National Park; Durban from Ukhlahlamba-Drakensberg Park; and Harare from Robert McIlwaine Recreational Park and Lake Robertson Recreational Park amongst several more. Many cities are set to increase their population by up to 85 per cent in the next fifteen years (Angel et al., 2011). Recognition of the importance of protected areas for water has not translated into solid policies for these new cities or even necessarily helped maintain existing links.

Chyulu Hills National Park in Kenya is part of the Tsavo-Amboseli wildlife complex, a series of low wooded hills that is also the source of most of the water for the rapidly growing port city of Mombasa 150 miles away on the coast. When the Zoological Society of London asked us to facilitate a management effectiveness assessment in 2015 we recorded a series of problems: illegal settlements including one that was going through the courts, illegal charcoal making and logging, industrial development isolating the park by cutting off migration routes and low management capacity. The fires from charcoal burning could be seen smudging the sky from miles away in Tsavo. The needs of local communities were clear, but further degradation would threaten the water supply for a whole city. Since then a variety of conservation projects have been developed in the area based on innovative funding mechanisms including carbon credits, lease payments for conservancy zones, watershed protection and ecotourism services, resulting in reduced overgrazing and water-intensive farming (Stolton and Dudley, 2019). The 7 million residents of Mombasa may know little or nothing about such projects, but they could make a huge difference to their quality of life.

This leads to a more general point: most large cities know where their water is coming from and have already made key decisions regarding its supply; the places where we need to focus attention over the next decade are the new cities springing up, particularly in Africa, China and other parts of Asia. A city of a million people can emerge in just a few years with the speed of modern building. The relationship of these new cities with hydrological cycles and natural ecosystems will determine the way that ecosystems are managed over major parts of the planet. It is on these medium cities that we should perhaps be focusing more of our attention.

Protected wetlands supply much more than just drinking water supplies. They are also critical to irrigation for agriculture, a far larger use of water overall, and a host of other provisioning and regulating services: nutrient recycling, supply of fish and other foods, building materials, flood regulation and cultural benefits. Inland fish catch in Africa exceeds 2.5 million tonnes a year (Humphrey, 2012). Wetlands absorb floodwater, protecting downstream communities (Pouget et al., 2013). Many protected wetlands attract tourists, for instance in Botswana the Okavango Delta brings 120,000 tourists/year, employing 600 guides. Standing wetlands are a key resource for the recharge of groundwater supplies, which in turn provide essential water supply during dry seasons or drought. Wetlands and particularly peatlands are globally important carbon stores and help to neutralise pollutants that would otherwise enter rural or urban drinking water supplies. Many wetlands have local sacred values; particularly lakes, springs and rivers such as the Ganges (WWF, 2009).

Food security

Water security is reasonably straightforward; taking the MEA typology it is a mixture of provisioning and regulating services focused everywhere on water *quality*, and in certain circumstances also on water *quantity*. The contribution of natural ecosystems to food security is much more complicated, involving many different elements and types of service. The Food and Agriculture Organisation of the United Nations has coined the term "biodiversity for food and agriculture" (BFA) to describe the multiplicity of ways in which ecosystem services support human diets. These range from species used directly for food and other products, species that support agriculture such as pollinators and crop wild relatives, and a variety of ecosystems supporting food and agriculture (FAO, 2019).

Particularly significant are wild plant and animal species used for food; species and habitats supporting food and agriculture (pollinators, crop and livestock wild relatives, soil organisms, predators and agricultural pest species and supportive habitats); and ecosystems supporting food production, including here the traditional farming and livestock systems found in many protected landscapes.

People have managed and collected wild resources for millennia, but today these resources have often been drastically reduced or even eliminated in the wake of expanding agricultural and other land uses. Protected areas are now often the only place they can be found but are rarely big enough to support large scale resource use from growing populations around their edges. Many protected areas manage this conundrum with permit controlled and sustainable collection of wild foods, so long as this does not undermine conservation objectives. As conservationists we have to say this is often not ideal, but rather a trade-off that builds enough local support for the protected area to survive. In the developed world this would generally be luxuries rather than staples, berries, mushrooms and the like; in the developing countries protected areas may provide access to fruit and wild vegetables that have disappeared elsewhere but are vital sources of nutrition.

Such approaches need management and conditions can change over time. A fashion for wild food in upmarket restaurants means that there is commercial over-collection of mushrooms and some other sought-after plants in the UK whereas this would not have been an issue ten years ago. On the morning before writing this section we were on our local estuary nature reserve collecting a small amount of samphire (*Crithmum maritimum*) for supper, which is allowed, and the wardens came to check we were not commercial collectors, which is not allowed and is now becoming a problem. In developing countries over-collection, including illegal collection, of wild plants and animals is a major threat in many protected areas, leading in extreme cases to what has become known as the "empty forest" effect.

The other major way in which protected and other conserved areas can contribute to wild food is by providing source populations that spill beyond their borders. This is best known in marine and to a lesser extent freshwater fishery. Marine protected areas (MPAs) provide safe havens for fish breeding and young fish; a proportion of the fish "spill over" beyond the MPA border, boosting fish stocks and preventing population collapse.

These ideas have been around for a long time. Twenty years ago, a review for WWF (Roberts and Hawkins, 2000) identified five main benefits of MPAs to fisheries. First, they enhance the production of offspring to restock fishing grounds. This is particularly by preserving older fish that produce a disproportionately large number of young; supporting sedentary species like clams that only reproduce successfully at high densities; and protecting vulnerable stages of the life cycle such as fish nurseries and spawning grounds. Second, fish spill over beyond the borders of MPAs and thus become available to fisherfolk with amounts influenced by factors such as the age of the reserve, continuity of habitat and mobility of the species. Then MPAs also allow some parts of the marine environment to escape the habitat damage that inevitably comes with trawling, anchoring and other human use. Fourth, they allow the development of natural marine communities, which may differ markedly from the culturally influenced communities in fishing grounds. And lastly, healthy natural ecosystems are more resilience against catastrophic events such as exceptional storms and the daily pressure from climate change.

There is now abundant evidence that MPAs can and do boost total fish stocks and thus support fisheries, including meta-studies of many different MPAs around the world (e.g. Gill, 2017). In some parts of the world this has long been recognised and practised by local communities. In the Pacific in particular such agreements have been in place for longer than any records exist and the term "locally managed marine areas" has emerged to describe these areas. Yet in other parts of the world, fishing communities regard such claims with deep distrust and routinely oppose marine protection of any kind; this has led to some of the bitterest disputes in the conservation arena.

Marine protected areas as local resources in the Philippines

The Island Garden City of Samal, popularly known as IGaCOS, is a kind of political sleight of hand; it is an island offshore of Davao City in Mindanao, the Philippines, that is classified legally as a "city" although in reality it is made up of scattered farms and villages in a predominantly rural area. Situated in the most politically troubled part of the Philippines, with a predominantly Moslem population in a Christian-dominated country, there have been terrorist incidents and kidnappings of foreigners, although it seemed very peaceful and friendly when I was there as part of a group working on a project to integrate climate change considerations into MPA planning (Belokurov et al., 2016). But the relevant point here is not about climate change strategies, but rather the reasons why we were able to be there and spend time working with the local community in the first place. The three small MPAs we were working on were not set up primarily by conservationists, but because local fisherfolk realised they needed to maintain fish stocks. In the shallow waters of Davao Bay, men literally stand on the bottom to set their nets, while in the background massive container ships sail past in the deeper channels between Samal and Mindanao. Fishing is critical to the economy of the island. And when we snorkelled past the root systems of some of the protected mangrove trees on offshore islands they were teeming with young fish, growing up protected from fishing pressure. Some of these fish will eventually stray out into the nets beyond the boundary of the reserve. Here conservation is not being driven primarily by outsider conservationists but by the pragmatic decisions of commercial fisherfolk. **ND**

Species that directly or indirectly support food and agriculture production are also important and often known as "associated biodiversity". Pollinating insects, birds and bats are essential for production on farms, gardens and orchards; increasingly these rely on protected areas to survive and spill over into farms in much the same way as seen in wild fish. Crops pollinated by animals account for 35 per cent of global food production (Klein et al., 2007). Bees are by far the most important pollinators, but wild insects, hummingbirds and bats all play a role (Garibaldi et al., 2013). Farmers in developed countries often pay beekeepers to place their hives near crops, with hives transported long distances from more pristine areas. Agriculture suffers when pesticides destroy insects; in parts of China farmers now have to pollinate their crops by hand (Partap and Ya, 2012).

Species that feed on the pests of crops and domesticated animals also support agriculture, either used deliberately through biological pest control methods or by maintaining healthy populations of prey-predators as in an organic farming system (Classen ct al., 2014).

Finally, crop wild relatives, species closely related to domesticated crops, are vital for crop breeding and adaptation such as drought and pest resistance (Maxted et al., 2007). Yet our own analysis showed that crop wild relatives are commonest in places with the lowest protected area coverage (Stolton et al., 2008a) and are a conservation priority (Castañeda-Álvarez et al., 2016). Livestock wild relatives have similar uses for livestock breeding although this is currently little utilised and their conservation status poorly understood (McGowan, 2010), suggesting the need for further protection (Redford and Dudley, 2018).

Finally, protected landscapes, IUCN protected area Category V, throughout the world support those traditional types of agriculture that would likely, without explicit support, disappear (Amend et al., 2008). We have already described the Potato Park in Peru. In these cases ecosystem services extend beyond support of food security, although this can be important, and enter the realms of food as cultural tradition: small vineyards, upland pastures rich with plants, free-range pig breeding in open woodland and terrace agriculture are all now increasingly confined to places where traditional farming is actively supported. In Spain for instance getting World Heritage status for the island of Mallorca has helped fund the renovation of many traditional terraces, while protected landscapes in Catalonia support traditional agriculture and vineyards. In Croatia, the Stari Grad Plain on the island of Hvav has remained practically intact since it was colonised by the Greeks in the fourth century BC and is producing grapes and olives today. The Japanese Satoyama initiative is highlighting such values. Such approaches are particularly important when rapid environmental change means that agronomists need to draw on as wide a range of crop varieties as possible to maximise adaptation potential. In coastal areas, MPAs often support traditional fisheries by banning large-scale commercial fishing.

Health benefits

Many issues discussed relating to food and water security have implications for health; childhood diarrhoea kills millions of children around the world and affordable supplies of clean water is the fastest way to reduce this toll. But protected areas support health more directly as well, by reducing the spread of some diseases, providing sources of local medicines and global pharmaceuticals, and perhaps most important of all by providing space for people to exercise both their minds and their bodies. Yet, like several of the other benefits described here they remain largely hidden to policy makers. A major study on links between biodiversity and health for the CBD and World Health Organisation in 2015 mentioned protected areas mainly as a potential "disbenefit" by reducing local populations' access to bushmeat (CBD and WHO, 2015).

Multiple studies have suggested that intact vegetation, particularly in forests, helps to reduce the rate of malaria infection, and that deforestation is associated with higher rates of infection (e.g., Oglethorpe, 2008). However, a recent macro-study in Africa did not confirm these findings and the authors speculated that the type and rate of forest loss may also influence malarial prevalence (Bauhoff and Busch, 2018). Conversely, it might be assumed that reducing wetland drainage through conservation in protected areas allows the survival of malarial mosquitoes and other water-borne diseases in these habitats.

Much less equivocal is the link between protected areas and both local and global medicines. The large majority of people in developing countries still rely primarily on medicines collected from the wild, a situation unlikely to change any time soon. For many people there is literally no alternative available or affordable (Hamilton et al., 2006). But this reliance runs deeper than lack of alternatives, even when people have funds to buy drugs to address acute health impacts they may go back to traditional medicines for chronic problems, love philtres and others (Welz et al., 2018). And while some traditional remedies provide little evidence of efficacy in clinical trials (rhino horn does not cure impotence) others are proving to have genuinely beneficial effects on health.

Some 88 per cent of UN Member States acknowledge use of herbal medicines through formal laws and policies (WHO, 2019). Yet today many popular medicinal herbs are declining or at risk of extinction; a recent study in the Himalayas found many important medicinal herbs were threatened (Tali et al., 2018). A combination of population growth coupled with loss of natural habitats is squeezing the resource, while commercial collectors are sometimes exhausting populations that would otherwise be sufficient for subsistence use. Managed extraction from protected areas is frequently filling critical gaps in supplies. A decade ago we identified a sample list of twenty protected areas around the world that were supplying from twenty to over 400 medicinal herbs to local communities (Stolton and Dudley, 2010); we know from anecdotal evidence that most protected areas in the tropics will be used as a resource of medicinal products, legally or illegally.

Medicinal herbs from protected areas in Bhutan

Cordyceps is a genus of fungus that is mainly parasitic on insects and popular in traditional Chinese medicine. In Bhutan, managing the *Cordyceps* harvest is a major task for protected area staff, but probably ensures support for conservation in the uplands. Protected areas are important in maintaining the sustainable management of resources such as *Cordyceps* to prevent outsiders from illegally profiting from resources controlled by the community. Mutually agreed management plans for non-timber forest products are a good tool to ensure sustainable resource use. Families are allocated certain areas in which to collect and maximum harvest levels, and rangers monitor the process and the populations of fungus. Most protected areas have some active management plans for collection. Community participation is encouraged in developing collection guidelines, and medical plant collection groups have been formed. We helped with a nationwide state of the parks survey in Bhutan a few years ago (Lham et al., 2019) and have sat with villagers discussing collection protocols; pragmatic agreements of this kind are increasingly being introduced into protected areas wherever medicinal plants remain an important resource. **SS/NS**

Perhaps even more important from an economic viewpoint is the use of protected areas as rich sites for the collection of animals, plants and microorganisms from which to develop commercial pharmaceuticals. Interest in use of wild species as the raw materials for development waxes and wanes; after a period of decline it seems to be booming again. In the past such materials were simply collected by explorer botanists and taken (sometimes smuggled) to their home countries; agreements like the CBD have slowed but not stopped this practice. More responsible companies today make agreements with countries that include financial compensation, in effect paying an exploration fee and gambling that the returns will more than repay the investment. Again, in the absence of alternatives protected areas are sometimes the preferred exploration sites. Famously, the thermophilic bacteria *Thermus aquaticus* was collected from hot springs in Yellowstone

National Park in 1966 and is used widely in medical compounds worldwide. We have recorded many other examples (Stolton and Dudley, 2010) including unfortunately many where the genetic material has been taken by the company without any recompense to the protected area.

Finally, but very importantly, protected areas provide places for people to take responsibility for their own mental and physical health by relaxation and physical exercise. The "green gym" concept recognises the role of nature reserves and other protected areas in providing safe and attractive places for people to walk, jog, ride bicycles and horses and take other forms of exercise. As non-communicable diseases become the major causes of premature death in the developed world, and obesity levels rocket upwards, it is probably the smaller reserves near urban centres that provide the most immediate resources. Keoladeo National Park in Rajasthan, India, is close to the major city of Bharatpur; local people enter the park before dawn to walk around the paths in the coolest part of the day. We have watched literally hundreds of people marching along briskly in the dark. And the majority of people visiting many nature reserves in the UK are now coming primarily to take a walk rather than identify rare species of birds; the paths and the presence of other people make them safe places in which to exercise.

Forests Europe identifies four ways in which forest areas contribute directly to health:

- Human health promotion and disease prevention: exercise, anti-stress programmes, mindfulness etc;
- Therapy and rehabilitation: programmes addressing approaches such as wilderness therapy, addressing social isolation;
- Education providing indirect health benefits: including various outdoor education programmes such as field trips, forest play groups etc; and
- Recreation and tourism: guided walks, educational trails, treetop walks, picnic sites etc (Marušáková and Sallmannshofer, 2019).

The Healthy Parks, Healthy People movement, originated by Parks Victoria in Australia, builds on these values to propose formal links between protected area agencies and institutions concerned with mental and physical health. Research shows that quite apart from the physical benefits of exercising in tranquil, beautiful places are therapeutic benefits for people with mental health problems and drug dependencies (Le Bas and Hall, 2008). A recent study determined the economic value of protected areas through improved mental health, using quality-adjusted life-years, is US$6 trillion a year, an order of magnitude greater than the tourism value (Buckley et al., 2019). The risks of nature deficit in children are increasingly being recognised. Parks Canada is current spearheading a global Nature for All campaign to bring more people around the world into contact with nature through protected areas (natureforall.global/). Health benefits from protected areas are multifactorial, complicated and still only starting to be understood.

Disaster risk reduction

In the build-up to the fifteen-year strategy of the UN International Strategy for Disaster Reduction (ISDR) IUCN ran an impressive advocacy for more use of natural ecosystems in defending against extreme climate events and earth movements. IUCN produced reports (Monty et al., 2016) and best practice guidelines (Dudley et al., 2015), assembled an impressive body of evidence and had words of support from the secretariat of the ISDR. Yet in the event governments virtually ignored natural solutions in the resulting Sendai Framework in 2015, except for one brief mention, way, way down the list of recommendations: "To strengthen the sustainable use and management of ecosystems and implement integrated environmental and natural resource management approaches that incorporate disaster risk reduction" (UN, 2015). So, while the use of ecosystem approaches to disaster risk reduction (DRR) continues to develop, it effectively does so outside the auspices of the UN body charged with global responses to disasters. The anecdote is worth mentioning in that it hints at the political inertia that ecosystem services often need to combat: generations of engineering companies have relied on DRR for their business, governments have their set approaches, and new ideas struggle for attention.

Natural ecosystems provide absorption and buffering against sudden movements of water, high winds, earth movements and slippage, and against the impacts of desertification and dust storms. Increasingly, given the rate of land use change, this means some form of area-based conservation to ensure that the DRR functions survive.

Natural flood plains absorb flood waters that would otherwise rush downstream and impact where people live; canalisation, drainage and levees have conversely increased downstream flooding on rivers like the Rhine in Europe. Similarly, vegetation on mountains and steep slopes provide a physical barrier and additional absorptive capacity when rain falls; hills denuded by deforestation and overgrazing let water slip down more quickly. Forested slopes protect against avalanches, rockfall and landslides; the periodic landslip disasters that hit the world headlines are almost always linked to deforestation. This can be particularly important in earthquake-prone areas; an earth movement can trigger a landslip and forests in Himalayan regions have been shown to reduce after-effects of a shake (Stolton et al., 2008b).

On the coast, natural barriers like mangroves, coral reefs, sand dunes and salt marshes play a similar role in providing buffers against storms and tsunamis, in some cases directly blocking water movement and in others providing space for it to dissipate harmlessly. In drylands, natural vegetation reduces wind erosion (and water erosion in periodic floods – dry areas suffer some of the world's worst flood events). This halts or slows the rate of desertification and, perhaps even more importantly from a human health perspective, cuts the vast dust storms that sweep across many desert areas and result in high rates of child breathing issues in wealthy countries like Kuwait.

We are summarising these benefits as if they are universally accepted, but in fact, this is not always the case. The science of DRR is complex and tricky in both ecosystem services and engineering solutions. The ISDR story at the start of this section is illustrative of a wider debate. Nothing works perfectly all the time, and huge disasters will overwhelm all defences.

So, for example, a flurry of research papers after the tsunamis in Sumatra, Indonesia and surrounding countries in 2004, and Japan in 2011 made claims and counter claims about the effectiveness of mangroves, coral reefs and other natural defences. Some said there was clear evidence of less damage and fewer deaths in places with healthy mangroves, while others argued that the evidence was ambiguous. The reality is that it probably depends; there is strong evidence that natural defences provide important protection in many cases although in a huge event they will, like everything else, fail. For instance, in Japan in the tsunami of 2011 whole trees were uprooted in places and added to the general mayhem. (People also stood on the defensive sea wall to watch the wave come in and were killed when it swept straight over.)

Avalanche control in Switzerland

The Parc Jura Vaudois in Switzerland follows the line of the Jura Mountains parallel with Lac Leman, opposite the Alps. Formed in 2010, it is a continuation and expansion of a regional park that has existed since the 1970s. It is rugged, limestone country with seasonal sheep and cattle farms and continuous cover forestry, looking down onto the lake and the vineyards and agriculture of the valley bottom. The Jura contains a remnant population of the European lynx (*Lynx lynx*), one of the most threatened cat species in the world, along with red (*Cervus elaphus*) and roe deer (*Capreolus capreolus*), chamois (*Rupicapra rupicapra*), wild boar (*Sus scropha*) and stone martins (*Martes foina*), plus a rich mountain flora. Rich too in cultural history, the annual practice of herding cattle back onto the mountains in the spring is now a major festival in towns like St Cergue. A tiny train winds up through the mountains and into France from the lakeside town of Nyon, and upland roads become ski tracks in the winter. We've explored the mountains in all seasons, on foot or snowshoes, and despite being a major playground for people from Geneva and Lausanne many areas remain remarkably quiet and unvisited. One of the reasons why the mountains are so heavily forested is the protection that they provide against avalanches; in all some 17 per cent of Swiss forests are classified for avalanche protection, with trees sometimes augmented by built structures to slow and disperse fast-moving snow. Like all defences, forests are not perfect; traces of past avalanches can be seen with cleared slopes covered with fast-growing mountain alder (*Alnus viridis*). But they are far better than nothing. **ND/SS**

Climate change strategies

Climate change presents natural resource managers with a unique set of challenges, although what that often means in practice is that it intensifies and further complicates existing challenges. Protected areas were originally designed to maintain landscapes and ecosystems; their role becomes more ambiguous when ecosystems themselves start to shift, species move, and climatic conditions stimulate rapid change. Addressing these issues provides a range of practical and also psychological or emotional challenges to anyone managing a protected area. But protected areas have a dual role here; undoubtedly victims of or challenged by climate change, but also tools that can help to mitigate and adapt to rapidly changing conditions. Two roles are important:

- **Mitigation:** through "storage", preventing the loss of carbon that is present in vegetation and soils; and "capture", sequestering carbon dioxide from the atmosphere in natural ecosystems.
- **Adaptation:** by maintaining ecosystem integrity, buffering local climate, reducing impacts from extreme events such as storms, droughts and sea-level rise and by maintaining essential ecosystem services that help people cope with changes in water supplies, fisheries, disease and agricultural productivity caused by climate change.

The role that protected areas play in adapting to climate change is in essence all the things described earlier – helping to maintain food and water security, buffering against disasters and so on – given an increased urgency by the added stress from climate change. Mitigation is another aspect and is basically the role of natural vegetation in maintaining and sequestering carbon dioxide and thus slowing, halting and reversing the rate of climate change.

After decades in which attention had been focused solely on fossil fuels (still by far the largest source of carbon dioxide), there is increasing recognition of the role of biomass burning and emissions from the drying of peat. Researchers calculated the loss from deforestation, with emissions often more serious from the soil below than from standing timber. Initial quick-fix solutions, such as plantations of fast growing trees, were found to offer few climate advantages and the importance of preventing land use change was highlighted. The UNEP World Conservation Monitoring Centre made a broad calculation that at least 15 per cent of the world's biomass was captured in protected areas. We worked with a series of organisations to highlight the role of protected areas in climate adaptation (Dudley et al., 2009), and marine biologists started to focus on the role of the ocean through carbon capture by microorganisms, kelp beds, seagrasses and the like (Laffoley and Grimsditch, 2009).

Carbon is captured by and builds up in vegetation and in the substrate below (soils, peat or the ocean floor). Contrary to earlier beliefs, this process does not cease once an ecosystem like a forest reaches a certain age, but old-growth forests, old seagrass beds and similar continue to build up carbon in the substrate below as plants and animals die and decompose (Baker et al., 2004). Catastrophic events, most importantly fire, can however result in a sudden release of carbon (and increased fire under climate change is one of the trickier problems facing those responsible for carbon management). Carbon storage below grassland and in agricultural soils is also important, giving traditional systems or those practices that consciously build soil health (organic farming for instance) an added advantage over those that treat the soil more like a holding substrate and end up with radically reduced carbon content (Lal, 2008). Area-based conservation, by keeping vegetation in place and soils undisturbed, and supporting traditional farming methods in protected landscapes, therefore acts to slow down the rate of climate change.

This thesis is not without its detractors. Climate change scientists and activists fear that this will provide governments with something easy to report on – simply publicising a "new" role for an existing protected area system – rather than addressing politically tricky issues like carbon taxes, energy reductions and the boom in flying created by cheap airlines. They point out that natural solutions are not enough on their own to address the climate change problem (Anderson et al., 2019). This is all true; the counter argument is that we need a whole range of responses, and to use all available options whilst making sure that governments do not get away with using one action as an excuse for ducking out of other responsibilities.

This does have wider implications for area-based conservation though, including particularly for the newer concepts like OECMs. The idea that we should be restoring forests on a far greater scale than previously is gaining momentum, with initiatives like the Bonn Challenge, a global effort to restore 150 million hectares of forests by 2020 and 350 million hectares by 2030; the Great Green Wall of Africa aiming to restore forests to slow down desertification; and others. Recently scientists calculated that a trillion more trees would be needed to offset climate change (Bastin et al., 2019), although these calculations have been challenged (Luedeling et al., 2019). All these initiatives look to the not-too-distant future when land becomes available because it is degraded and no longer suitable for agriculture, or because technological changes (e.g. meat substitutes) free up land.

Restoration is an important process in many protected areas, particularly if they have undergone degradation before they were set up, or due to failings in management effectiveness. We'll talk about this more later. But restoration within a protected area needs to follow best practices to ensure that it fits the conservation objectives (Keenleyside et al., 2012); in other words, ecological restoration rather than simple tree-planting. Restoration on the scale being suggested would almost certainly involve tree planting of a rate that would not fit with full ecological restoration and might provide some conservation benefits but certainly not a full range. Such areas might need to be modified over time to gain more natural characteristics (e.g., stepwise restoration, Dudley and Maginnis, 2018) or fall outside the conservation estate. Given that natural grasslands are arguably declining more quickly than forests at the moment this also creates risks that massive tree planting could be antithetical to wider conservation aims. Integrating ecosystem-based climate mitigation with agriculture, food security and biodiversity conservation will require planning and trade-offs that have scarcely started to be discussed.

Recreation and tourism

Tourism is now probably the number one industry in the world, depending on how the figures are calculated, and it is the political and economic force that maintains many protected areas. We will make an educated assertion that on a day-to-day basis, most rangers in most popular protected areas spend more time looking after tourism needs than in active conservation. Many years ago, we worked in the iconic Serengeti as part of a World Heritage project on management effectiveness. At that time tourism activities were the focus of 50 per cent of park operations, and the park depended on tourism for income. Our work was with the two park ecologists tasked with maintaining the roughly 14,750 km^2 national park, whilst most of the other fourteen senior staff worked on tourism. The World Heritage committee has consistently had concerns over the tourism developments in the park, the most recent being a proposed new lodge within the "low-use zone" close to a wildebeest migration route.

Protected areas have an uneasy relationship with tourism. Many countries, companies and individuals set up protected areas, and equally importantly maintain those areas, because of their value in bringing in tourists. But management in some of our most iconic protected areas is being dominated by tourism interest, often in conflict with conservation management. We visited the Galapagos Islands during an economic downturn when there were apparently comparatively few visitors yet were still amazed by the density of tour boats, and visitor quotas have been increased again since. Some of these places are literally being loved to death. Others limit tourist numbers by focusing on high-end tourism, but in the process confine wildlife experiences to a moneyed elite.

Tourism is also one of the major drivers of climate change through its extraordinary impact on flying, creating a massive paradox at the heart of many protected area strategies. On the one hand, protected areas survive, particularly in poorer countries, largely due to the fact that they bring in foreign exchange. This also boosts political support for conservation around the world; research shows that once people have seen iconic wildlife close at hand, they feel more concerned with its survival. But everyone with an environmental conscience knows we should be cutting out extraneous air trips and that we may well be the last generation to travel so much, meaning that tourism strategies based on foreign visitors may fail in the future. This will be fine for countries like South Korea or Japan where the nature-based tourism sector is overwhelmingly domestic, but it bodes ill for those developing countries where virtually all tourists come from abroad. Building a domestic market for ecotourism is a critical conservation strategy in many countries of Africa and Asia for example.

For now, recreational use of protected areas takes a number of forms. Day visits to nature reserves and national parks by domestic recreational users or holidaymakers is important and may be a dominant use in smaller reserves or in national parks near centres of population. Many of the visitors who come there will be attracted to nature in general terms, possibly through landscape values or a desire to spend time in a peaceful rural environment, and a minority will be interested in particular groups, particularly birds, butterflies and wildflowers. Footpaths, cycle trails, viewing platforms, bird hides and places to have picnics are all needed to help cater for these visitors, along with interpretation material to help explain what is going on and what people might expect to see.

Making the best economic use of remaining forests in Rwanda

Rwanda underwent a catastrophic civil war in 1994, resulting in up to a million people being killed in what was widely seen as genocide. The country has worked hard to rebuild its economy, society and global standing. Rwanda is small and crowded with rich, fertile soils and most available land is given over to agriculture with terraces on steep slopes and everywhere stands of alien *Eucalyptus*. Yet the government has prioritised its national park system, largely as a vehicle for attracting foreign tourists. Gorilla tourism has been popular in the country since the 1950s but virtually disappeared between 1994 and 1998 during the war and subsequent instability. Since then it has boomed. By 2008, there were 20,000 protected area visits of which 17,000 were for gorilla viewing (Nielsen and Spenceley, 2010). Growth has continued, rising 30 per cent between 2014 and 2016. Tourism earned Rwanda US$400 million in 2016 (Hepola, 2017) and US$438 million in 2017 (Mwai, 2018), making it the largest earner of foreign exchange. There are seven mountain gorilla groups habituated in Volcanoes National Park, making 56 gorilla permits for visitors available every day at a cost of US$1,500 each. The country is intentionally targeting high-end tourism. There is a revenue sharing strategy, which reinvests 5 per cent of earnings back to local communities, but this only represents a small amount per head and seems to miss many of the poorest communities (Munanura et al., 2016). Supported infrastructure projects like water tanks and buffalo walls are perceived as more successful than income generating projects (Bush et al., 2010). Research suggests that economic benefits have not substantially trickled down to the local communities and tensions (including poaching) remain (Sabuhoro et al., 2017). Local people get no discount on gorilla watching. Ecotourism is therefore generating very important economic benefits, but further work is needed to ensure equitable distribution of benefits. **SS/ND**

Tourism in larger, wilder or more dangerous protected areas like national parks and wilderness areas is both more exclusive (mainly due to the costs of reaching and going into such protected areas) and more specialised. Guided tourism has increased dramatically in the last few decades and more visitors now take part in organised activities in groups than was the case in the past, which is generally good for local employment and thus community responses to conservation strategies. Countries vary in the way in which access is granted; generally freely available in national parks in Europe, subject to fees in many other places and with numbers of visitors strictly controlled in some areas. The canoe trip we took with friends through the Killarney Lakes in Ontario, Canada, was reserved on the day that booking opened in the spring and we still didn't get all the camping places we wanted.

There are no easy answers, and we recognise that we are also part of the problem; although we have been lucky enough to have the chance to visit many protected areas professionally, we have also taken the opportunity to visit many more on our travels. We believe the doctors and psychologists who say that contact with nature is critical to human mental and physical health and for many people this translates into trips to protected areas. But many national parks have priced themselves into a category where only the comparatively wealthy can take advantage of the benefits. There is clearly an unresolved anomaly here.

Cultural and spiritual values

Most of the world's faiths and religions have a close connection to nature. This is an integral part of the faith in many small animist belief systems and to a more subtle extent in the great religions arising in Asia: Hinduism and its offshoots, Taoism, Buddhism, Shinto in Japan and so on. But even amongst the three monotheistic faiths arising in the Middle East, Judaism, Christianity and Islam, where nature worship was regarded as a form of idolatry and for instance the old sacred sites, such as those of the Druid religion, were often systematically destroyed, a deep reverence for nature remains and certain natural areas gain importance in consequence.

There are three main way in which faiths interact directly with protected areas: through the existence of sacred natural sites (SNS), or because religious buildings and communities exist in protected areas, or because pilgrimage routes or other active forms of worship take place within national parks and nature reserves (Dudley et al., 2009).

SNS are areas that have attained particular importance to a faith group, large or small. They include woodland, individual trees, springs, lakes, waterfalls, rivers, caves, mountains and areas with distinctive features, like large boulders or geological oddities that attract stories and legends. Many are intensely local, such as a sacred grove or spring recognised by a single village. Others span whole continents, such as the "songlines" and innumerable sacred sites criss-crossing Australia. Some places have sacred importance to multiple groups, like Mount Kailash in the Tibetan autonomous region of China, which is sacred to Jains, Hindus, Buddhists and those following the Bon religion. Many are old, although by no means all, and new SNS emerge all the time. When we visited a site in Madagascar where archaeologists had found evidence of early landings and settlement there were also offerings there; local people had seen the archaeological activity and recognised something significant had happened there and incorporated the area into their local pattern of worship.

The importance of SNS has been increasingly recognised in recent years, both from the perspective of their cultural and spiritual significance to many communities and due to the fact that many are strictly protected and thus have important conservation values as well (Dudley et al., 2010). SNS have proven remarkably resilient in surviving transition from one faith to another; for instance, they often continue to be respected, sometimes surreptitiously, in communities that have converted to Islam or Christianity many years previously, or in hard-line communist countries where all religion is suppressed.

Larger religions interact with protected areas as well. Rila Monastery is believed to be the oldest monastery in the Slav world and the largest active religious centre in Bulgaria. The current buildings were completed in the fifteenth century after an earlier construction was destroyed by fire. A major focus for pilgrimage, particularly during the long Ottoman occupation, the monastery was inscribed onto the World Heritage list in 1983. Situated at the north-western end of the Rodolphi Mountains, it is also within Rila National Park, an 800 km^2 area designated in 1992 to protect mountain ecosystems. The protected area is the source of many rivers and contains 120 glacial lakes, while a range of rare species make it one of the most important protected areas in Europe. The monks have been actively involved in conservation activities within their own lands in the national park.

Protection for people

Finally, protected areas protect people, though not everywhere, and maybe not even most of the time; we will be looking at some depressing stories of expulsions and human rights abuses in the name of conservation in the next chapter. But over the past few decades, and particularly since the fifth World Parks Congress in Durban in 2003, the role of protected areas in also protecting vulnerable human communities has been increasingly recognised.

When Arthur Conan Doyle, the creator of Sherlock Holmes, wrote his novel *The Lost World*, he based it around the idea that there were still mountain plateaus high above tropical forests so remote that humans had literally never been there. He imagined a world where Professor Challenger and his friends discovered dinosaurs continuing to live into the nineteenth century, a concept that has spawned a hundred monster movies. While discovering giant dinosaurs now seems unlikely, the idea of unclimbed, forest-covered plateaus cut off by sheer cliffs is absolutely believable. We have been lucky enough to see some of them at close quarters, flying in a small plane over Chiribiquete National Park, a natural World Heritage site in Colombia. *Tepui* is an Arawak Indian word for large, rocky, table-topped plateaus found in the part of the Amazon within the Guiana Shield, whose irregular shapes and vertical walls make them virtually unique (Sánchez-Castillo et al., 2014). Most of those mountains have never been climbed, and we would have given a lot to be able to explore those forests on foot. But we can't. An un-contacted group of indigenous people live there and by their actions have made it clear that they wish to be left alone. As part of a far-sighted policy towards indigenous rights in Colombia, the government has decreed that the protected area remains off limits. Even our colleagues who undertook a scientific expedition there to collect information for the World Heritage nomination had to turn back when they saw signs of human presence. Chiribiquete is one of a number of such protected areas in Latin America; on the one hand set aside to protect near-intact forest but also kept strictly off limits to allow people to continue traditional lifestyles without harassment.

Such example are fairly extreme cases, and only really feasible in countries where there is still large areas of intact forest and still indigenous people voluntarily remote from the rest of humanity. But other models are becoming increasingly common. In Colombia again, the Sanctuary of Flora Medicinal Plants Orito Ingi Ande was proposed by the indigenous Kofán communities and medicine-men of the Putumayo foothills, as part of their strategy to strengthen and restore their traditional culture. It is jointly managed with the national parks agency. In Canada, local First Nation groups supported expansion of the Nahanni National Park in part to maintain traditional cultures. We have already mentioned the large network on Indigenous Protected Areas in Australia. In many places in upland Europe, traditional transhumance activities only survive because they are in protected areas.

The boreal Laponian area of Sweden is the summer home of about 1,140 Sámi people in nine active reindeer-herding communities (*samebys*) (Reimerson, 2016), who use the area to graze 65,000 reindeer (Svels and Sande, 2016). It is globally significant as an area where an ancestral way of life is based on the seasonal movement of reindeer and for that reason is recognised as a World Heritage site. The site is 99 per cent state-owned and composed of four national parks (Stora Sjofallet, Sarek, Padjelanta and Muddus) and two nature reserves (Saunja and Stubba). The Sámi culture is based on the concept of "maintenance of life" (*birgejupmi*), which combines people, natural resources, spiritual and psychological health, and implies a close connection between landscape, environment and ecosystems (Porsanger, 2012). The continued survival of the cultural and natural environment of Laponia can be credited in great part to a management approach that has attempted to integrate conservation with Sámi herder perspectives, although reaching this situation took fifteen years of negotiation and some conflict (Reimerson, 2016). This led to a collaborative management structure being established in 2011. Since then, the non-profit, locally based "Laponiatjuottjudus Association", made up of five representatives from Sámi reindeer herding communities and four protected area representatives from the state, region and two municipalities, has managed the area according to a management plan (Tjuottjudusplána, 2012).

The role of protected areas in maintaining traditional livelihoods is a critical part of management in almost all protected landscapes and in many wilderness areas and other reserves that rely on traditional management. These aspects are still often missed by many observers.

One particular aspect of the social role of protected areas is the recognition of "peace parks": protected areas that cross regional or national boundaries (transboundary protected areas) where there have been wars and long-running disputes and where the protected area becomes a neutral ground for reconciliation and peace-building (Ali, 2007). The first official peace park was recognised in 1932, the Waterton-Glacier International Peace Park between the United States and Canada, itself a recognition of a long-standing peace rather than a means of conflict resolution. Other examples have sought to defuse conflicts, such as the peace park between Peru and Ecuador in the Cordillera del Condor region (Kakabadse et al., 2016), while proposals exist in places with relatively recent conflict like the Balkans (Walters, 2015) and Korea (Kim and Bueno de Mesquita, 2015).

The implications of ecosystem services for management of area-based conservation

There is strong evidence that area-based conservation is highly beneficial, and probably essential, to the delivery of many ecosystem services. However there is still a lot to be worked out about what this means in practice.

Protected areas produce a wide range of ecosystem services. Some do so consciously, with clear integration of ecosystem service delivery into conservation management plans and with a working relationship and support, including sometimes financial support, from the recipients. In other cases, these ecosystem services exist virtually unrecognised and provide no active support to the protected area. A third group of services is at best tolerated and sometimes actively detrimental to wider conservation objectives.

So, for example water services, disaster risk reduction, protection of crop wild relatives, maintenance of many but not all spiritual values and storage and sequestration of carbon can all usually be utilised without interfering with or undermining conservation in any way. An exception here might be if carbon sequestration took precedence over conservation, for instance by afforestation on natural grassland; although noted as a potential problem there are few signs of this happening. If we include tourism amongst ecosystem services the picture becomes slightly more complicated; tourism helps build public support for conservation by showing people beautiful landscapes and seascapes and interesting wildlife, but it also undoubtedly adds to the pressure facing many protected areas.

Other ecosystem services are much more of a compromise: fishing, hunting, collection of wild foods and medicines, fuelwood and timber extraction, grazing and fodder collection are integrated into management with varying degrees of success but are almost always a trade-off with nature conservation. An exception to this is for cultural landscapes where present biodiversity relies to some extent on traditional cultural management although this itself can lead into complicated debates about the deeper aims of conservation. These trade-offs may make perfect sense in terms of human rights and social values and we would generally support such management decisions along with their inevitable complications, debates, trade-offs and tensions.

It is also worth noting that some of these more problematic trade-offs may disappear in the normal course of social development. We visited one of the remote offshore islands in Dadohaehaesang National Park in South Korea a few years ago, a couple of hours away from the mainland in a fast boat, where the Korea National Park Service was seeking advice about the fact that residents were still collecting shellfish from coastal pools. Indeed, they were; clusters of people were scrabbling around on the shore collecting seafood. But all the remaining people were old, many very old; all the younger people had left for the mainland in a social movement almost certainly unconnected with the national park establishment. The minimal impact of local inhabitants living out their lives in the place they want to stay is perfectly justifiable.

It would however be naïve to assume that extraction of natural resources from protected areas will simply fade away as living conditions change. In South Africa most people now use Western medicines when they are seriously ill but still rely on medicines from native vegetation for love potions and similar, with rising living standards boosting demand. Rising incomes in China and Viet Nam have led to an alarming rise in the wildlife trade, with tiger meat now said to have a street value in excess of an equivalent weight of heroin. Bushmeat hunting in national parks in many parts of the world is no longer driven primarily by subsistence needs but to feed the tastes of an aspiring urban middle class. These pressures cannot be addressed by substitution or increasing rural incomes, but only by a mixture of long-term education and more effective enforcement activities.

Interventions by protected area managers or by wider planning processes can reduce some pressures. Around Nyungwe National Park in Rwanda local planning has encouraged establishment of eucalyptus plantations to provide an alternative source of fuelwood and building materials; along with tea plantations that create a buffer for the natural forest, the plantations are themselves rich in bird life and provide sustainable income for local people. The flowering of the eucalyptus provides rich sources of pollen and a proportion of the park fees has helped set up a local beekeeping collective; we watched them having a meeting when we visited recently and a plantation of eucalyptus is a good swap for a reduction in fuelwood collection in the native forest.

In some ways protected areas provide ideal conditions for exploitation of ecosystem services: they have set boundaries, management plans, trained staff and equipment, legal or other effective forms of protection and management aims that are usually compatible with ecosystem service delivery. But not everyone agrees. Some conservationists argue strongly that mixing biodiversity protection and ecosystem service delivery simply risks confusion and that protected areas should focus on conservation, leaving ecosystem services to other area-based conservation initiatives such as OECMs. They have a point and importantly ecosystem services offer opportunities to set land and water aside in places where protected areas would be hard to establish. Carbon projects in sub-Saharan Africa have to date centred on areas specifically "unsuitable" for establishment of classical African national parks, because of their terrain, or infestation with tse tse fly (*Glossina* spp.), or for other reasons, thus in effect adding to the conservation estate without adding to the protected area estate.

On the other hand, many protected areas survive because of the additional social and political support provided because their ecosystem services are recognised and valued. It would be perverse to deny these values with an ideological argument that protected areas should "only" protect biodiversity. And protected areas can sometimes benefit from financial incentives as well, such as Payment for Ecosystem Service schemes (see Chapter 10).

In the next few years, four urgent steps are needed to consolidate the role of area-based conservation in conservation and delivery of ecosystem services. First, we need a much more thorough understanding than we have at the present about the true extent and value of these services around the world. Second, more sophisticated tools and approaches are needed to understand, negotiate and where necessary painlessly phase out damaging resource extraction within protected areas. At a broader scale, this needs to extend to an understanding of exactly where ecosystem services fit into the broader approaches to area-based conservation that are now emerging around the world. And lastly, we need to learn from the successful efforts of entrepreneurs in converting these services into hard cash and other benefits that can help to support the people who maintain them.

References

Ali, S.H. (ed.) (2007) *Peace Parks: Conservation and Conflict Resolution*. MIT Press, Cambridge, MA

Amend, T., Brown, J., Kothari, A., Philips, A. and Stolton, S. (eds.) (2008) *Values of Protected Landscapes and Seascapes Volume 1: Protected Landscapes and Agrobiodiversity Values*. GIZ and IUCN, Eschborn, Germany and Gland, Switzerland

Anderson, C., De Fries, R.S., Litterman, R., Matson, P.A., Nepstad, D.C., et al. (2019) 'Natural climate solutions are not enough'. *Science*, vol. 363, no. 6430, pp 933–934

Angel, S., Parent, J., Civco, D.L. and Blei, A.M. (2011) *Making Room for a Planet of Cities*. Lincoln Institute of Land Policy, Cambridge

Baker, T.R., Phillips, O.L., Malhi, Y., Almeida, S., Arroyo, L., et al. (2004) 'Increasing biomass in Amazon forest plots'. *Philosophical Transactions of the Royal Society B*, vol. 359, pp 353–365

Bastin, J.F., Finegold, Y., Garcia, C., Mollicone, D., Rezende, M., et al. (2019) 'The global tree restoration potential'. *Science*, vol. 365, pp 76–79

Bauhoff, S. and Busch, J. (2018) *Does Deforestation Increase Malaria Prevalence? Evidence from Satellite Data and Health Surveys*. Working Paper 480, Centre for Global Development

Belokurov, A., Baskinas, L.T., Biyo, R., Clausen, A., Dudley, N., et al. (2016) *Climate Adaptation Methodology for Protected Areas (CAMPA) Coastal and Marine*. WWF, Gland, Switzerland

Bhandari, B.B., Oh Suh, S. and Woo, S.-H. (eds.) (2008) *Water Towers of Asia: Experiences in Wetland Conservation in Nepal*. IUCN Nepal and Gyeongnam Ramsar Environmental Foundation South Korea, Kathmandu and Gyeongnam

Buckley, R., Brough, P., Lague, L., Chauvenet, A., Fleming, C., et al. (2019) 'Economic value of protected areas via visitor mental health'. *Nature Communications*, vol. 10, no. 5005. Doi:10.1038/s41467-019-12631-6

Bush, G.K., Ikirezi, M., Daconto, G., Gray, M. and Fawcett, K. (2010) *Assessing Impacts from Community Conservation Interventions Around Parc National des Volcans, Rwanda*. Dian Fossey Gorilla Fund International, CARE International and International Gorilla Conservation Programme

Castañeda-Álvarez, N.P., Khoury, C.K., Achicanoy, H.A., Bernau, V., Dempewolf, H., et al. (2016) 'Global conservation priorities for crop wild relatives'. *Nature Plants*, vol. 2. Doi:10.1038/NPLANTS.2016.22

Caufield, C. (1991) 'A reporter at large: The ancient forest'. *The New Yorker*, 14 May 1990

CBD and WHO. (2015) *Connecting Global Priorities: Biodiversity and Human Health*. CBD Secretariat, Montreal

Classen, A., Peters, M.K., Ferger, S.W., Helbig-Bonitz, M., Schmack, J.M., et al. (2014) 'Complementary ecosystem services provided by pest predators and pollinnatoirs increase quantity and quality of coffee yields'. *Proceedings of the Royal Society B*, vol. 281, Article Id. 20133148

Dudley, N., Bhagwat, S., Higgins-Zogib, L., Lassen, B., Verschuuren, B. and Wild, R. (2010) 'Conservation of biodiversity in sacred natural sites in Asia and Africa: A review of the scientific literature'. In: Verschuuren, B., Wild, R., McNeely, J. and Oviedo, G. (eds.) *Sacred Natural Sites: Conserving Nature and Culture*. Earthscan, London, pp 19–32

Dudley, N., Higgins-Zogib, L. and Mansourian, S. (2009) 'The links between protected areas, faiths, and sacred natural sites'. *Conservation Biology*, vol. 23, no. 3, pp 568–577

Dudley, N., Hockings, M. and Verschuuren, B. (2015) 'To go, or not to go? What are the business attitudes to the philosophy of no-go policies and protected areas?' *PARKS*, vol. 21, no. 2, pp 7–10

Dudley, N. and Maginnis, S. (2018) 'A stepwise approach to increasing ecological integrity in forest landscape restoration'. *Ecological Restoration*, vol. 36, no. 3, pp 174–176

Dudley, N. and Stolton, S. (eds.) (2003) *Running Pure: The Importance of Forest Protected Areas to Drinking Water*. WWF International and The World Bank, Gland, Switzerland and Washington, DC

FAO. (2019) *The State of the World's Biodiversity for Food and Agriculture*, Bélanger, J. and Pilling, D. (eds.). FAO Commission on Genetic Resources for Food and Agriculture Assessments, Rome

Garibaldi, L.A., Steffan-Dewenter, I., Winfree, R., Aizen, M.A., Bommarco, R., et al. (2013) 'Wild pollinators enhance fruit set of crops regardless of honey bee abundance'. *Science*, vol. 340, pp 1608–1611

Gill, D.A., Mascia, M.B., Ahmadia, G.N., Glew, L., Lester, S.E., et al. (2017) 'Capacity shortfalls hinder the performance of marine protected areas globally'. *Nature*, vol. 543, pp 665–669

Hamilton, A., Dürbeck, K. and Lawrence, A. (2006) 'Towards a sustainable herbal harvest: A work in hand'. *Plant Talk*, vol. 43, January

Hamilton, L., Dudley, N., Greminger, G., Hassan, N., Lamb, D., et al. (contributors) (2008) *Forests and Water*. FAO Forestry Paper 155. Food and Agricultural Organization, Rome

Hepola, S. (2017) 'How Rwanda became the unlikeliest tourism destination in Africa'. *Bloomberg Businessweek*, 28 September 2017. www.bloomberg.com/news/features/2017-09-28/how-rwanda-became-the-unlikeliest-tourism-destination-in-africa. Accessed 3 September 2019

Humphrey, S. (2012) *Africa's Ecological Footprint*. WWF and ADB

Kakabadse, Y., Caillaux, J. and Dumas, J. (2016) 'The Peru and Ecuador peace park: One decade after the peace settlement'. In: Bruch, C., Muffett, C. and Nichols, S.S. (eds.) *Governance, Natural Resources and Post-Conflict Peacebuilding*. Taylor & Francis, London

Keenleyside, K., Dudley, N., Cairns, S., Hall, C. and Stolton, S. (eds.) (2012) *Ecological Restoration for Protected Areas: Principles, Guidelines and Best Practice*, Best Practice Protected Area Guidelines no. 18. IUCN, Gland, Switzerland

Kim, C.H. and Bueno de Mesquita, B. (2015) 'Ecological security and the promotion of peace: A DMZ eco-peace park'. *Korean Journal of Defence Analysis*, vol. 27, pp 539–557

Klein, A.-M., Vaissiere, B.E., Cane, J.H., Steffan-Dewenter, I., Cunningham, S.A., et al. (2007) 'Importance of pollinators in changing landscapes for world crops'. *Proceedings of the Royal Society B: Biological Sciences*, vol. 274, pp 303–313

Laffoley, D. and Grimsditch, G. (2009) *The Management of Natural Coastal Carbon Sinks*. IUCN, Gland, Switzerland

Lal, R. (2008) 'Carbon management in agricultural soils'. *Mitigation and Adaptation Strategies for Global Change*, vol. 12, pp 303–322

Larsen, F.W., Turner, W.R. and Mittermeier, R.A. (2014) 'Will protection of 17% of land by 2020 be enough to safeguard biodiversity and critical ecosystem services?' *Oryx*, vol. 49, no. 1, pp 74–79

Le Bas, B. and Hall, J. (2008) 'Conservation therapy – hands-on examples from national nature reserves'. *Ecos*, vol. 29, p 2

Lham, D., Wangchuk, S., Stolton, S. and Dudley, N. (2019) 'Assessing the effectiveness of a protected area network: A case study of Bhutan'. *Oryx*, vol. 53, no. 1, pp 63–70

Luedeling, E., Börner, J., Amelung, W., Schiffers, K., Shepherd, K. and Rosenstock, T. (2019) 'Forest restoration: Overlooked constraints'. *Science*, vol. 366, no. 6463

Marušáková, L. and Sallmannshofer, M. (eds.) (2019) *Human Health and Sustainable Forest Management*. Forests Europe, Zvolen, Slovak Republic

Maxted, N., Ford-Lloyd, B.V. and Kell, S.P. (2007) 'Crop wild relatives: Establishing the context'. In: Maxted, N., Ford-Lloyd, B.V., Kell, S.P., Iriondo, J., Dulloo, E. and Turok, J. (eds.) *Crop Wild Relative Conservation and Use*. CABI Publishing, Wallingford, pp 3–30

McGowan, P.J.K. (2010) 'Conservation status of wild relatives of animals used for food'. *Animal Genetic Resources*, vol. 47, pp 115–118

Monty, F., Murti, R. and Furuta, N. (2016) *Helping Nature Help Us: Transforming Disaster Risk Reduction Through Ecosystem Management*. IUCN, Gland, Switzerland

Munanura, I.E., Backman, K.F., Hallo, J.C. and Powell, R.B. (2016) 'Perceptions of tourism revenue sharing impacts on volcanoes national park, Rwanda: A sustainable livelihoods framework'. *Journal of Sustainable Tourism*. http://dx.doi.org/10.1080/09669582.2016.1145228

Mwai, C. (2018) 'Rwanda: Tourism rakes in $438 million in 2017'. *The New Times*, 28 August. https://allafrica.com/stories/201808290251.html. Accessed 3 September 2019

Nielsen, H. and Spenceley, A. (2010) *The Success of Tourism in Rwanda – Gorillas and More*. Paper for the World Bank and SNV, Washington, DC

Nobre, A.D. (2014) *The Future Climate of Amazonia: Scientific Assessment Report*. ARA, CCST-INPE, INPA. São José dos Campos, SP, Brazil

Oglethorpe, J., Honzak, C. and Margoluis, C. (2008) *Healthy People, Healthy Ecosystems: A Manual for Integrating Health and Family Planning into Conservation Projects*, World Wildlife Fund, Washington, DC

Partap, U. and Ya, T. (2012) 'The human pollinators of fruit crops in Maoxian County, Sichuan, China: A case study of the failure of pollination services and farmers' adaptation strategies'. *Mountain Research and Development*, vol. 32, pp 176–186

Porsanger, J. (2012) 'Indigenous Sámi religion: General considerations about relationships'. In: Mallarach, J.-M., Papayannis, T. and Väisänen, R. (eds.) *The Diversity of Sacred Lands in Europe: Proceedings of the Third Workshop of the Delos Initiative – Inari/Aanaar 2010*. IUCN and Vantaa, Gland, Switzerland and Metsähallitus Natural Heritage Services, Finland

Pouget, C., Jean-Baptiste, N. and Kabisch, S. (2013) *Responding to Climate Change in Africa*. CLUVA, Leipzig

Redford, K.H. and Dudley, N. (2018) 'Why should we save the wild relatives of domesticated animals?' *Oryx*, vol. 52, pp 397–398

Reimerson, E. (2016) 'Sámi space for agency in the management of the Laponia world heritage site'. *Local Environment*, vol. 21, pp 808–826. Doi:10.1080/13549839.2015.1032230

Roberts, C.M. and Hawkins, J.P. (2000) *Fully-Protected Marine Reserves: A Guide*, WWF Endangered Seas Campaign, Washington, DC and Environment Department, University of York

Sabuhoro, E., Wright, B., Munanura, I.E., Nyonza Nyakabwa, I. and Nibigira, C. (2017) 'The potential of ecotourism to generate support for mountain gorilla conservation among local communities neighbouring volcanoes national park, Rwanda'. *Journal of Ecotourism.* Doi:10.1080/14724049.2017.1280043

Sánchez Castillo, V., Alzate, M.P., Castaño Uribe, C., Willingshoffer, A. and Marmon, T. (2014) *Parque Nacional Natural Chiribiquete Colombia.* GIZ, Bogotá

Satterthwaite, D. (2014) *Cities of More Than 500,000 People, Visualisation.* International Institute for Environment and Development, London. www.iied.org/cities-interactive-data-visual. Accessed 8 November 2019

Stolton, S., Boucher, T., Dudley, N., Hoekstra, J., Maxted, N., et al. (2008a) 'Ecoregions with crop wild relatives are less well protected'. *Biodiversity*, vol. 9, pp 52–55

Stolton, S. and Dudley, N. (2010) *Vital Sites: The Contribution of Protected Areas to Human Health.* WWF, Gland, Switzerland

Stolton, S. and Dudley, N. (2019) *The 'New' Lion Economy: Unlocking the Value of Lions and Their Landscapes.* Equilibrium Research, Bristol

Stolton, S., Dudley, N. and Randall, J. (2008b) *Natural Security: Protected Areas and Hazard Mitigation,* WWF International, Gland, Switzerland

Svels, K. and Sande, A. (2016) 'Solving landscape-related conflicts through transnational learning? The case of transboundary Nordic world heritage sites'. *Landscape Research*, vol. 41, pp 524–537. Doi:10.1080/01426397.2016.1151485

Tali, B.A., Khuroo, A.A., Nawchoo, I.A. and Ganie, A.H. (2018) 'Prioritising conservation of medicinal flora in the Himalayan biodiversity hotspot: An integrated ecological and socioeconomic approach'. *Environmental Conservation.* Doi:10.1017/S0376892918000425

Tjuottjudusplána. (2012) *The Management Plan of the Laponian World Heritage Site.* https://laponia.nu/wp-content/uploads/2014/08/Laponia-forvaltningsplan-eng-web-150327_2.pdf. Accessed 26 March 2018

UN. (2015) Resolution Adopted by the General Assembly on 3 June, 69/283 Sendai Framework for Disaster Risk Reduction 2015–2030

UNCCD. (2017) *Global Land Outlook.* UN Convention to Combat Desertification, Bonn, Germany

Walters, J.T. (2015) 'A peace park in the Balkans: Cross-border cooperation and livelihood creation through coordinated environmental conservation'. In: Young, H. and Goldman, L. (eds.) *Livelihoods, Natural Resources, and Post-Conflict Peacebuilding.* Earthscan, London, pp 155–166

Welz, A.N., Emberger-Klein, A. and Menrad, K. (2018) 'Why people use herbal medicine: Insights from a focus-group study in Germany'. *BMC Complementary and Alternative Medicine*, vol. 18, p 92. Doi:10.1186/s12906-018-2160-6

WHO. (2019) *WHO Global Report on Traditional and Complementary Medicine 2019.* World Health Organization, Geneva

WWF. (2009) *Sacred Waters: Cultural Values of Himalayan Wetlands.* WWF, Nepal, Kathmandu

10

EFFECTIVENESS OF THE EXISTING ESTATE

Does area-based conservation deliver? And if so, what does it deliver and at what cost? Why is biodiversity declining in a period of rapid protected area expansion? Do top-down state-controlled protected areas function better or worse than those set up and managed by communities? Protected areas face a bewildering array of threats; are these being mitigated? Over the past two decades, interest in management effectiveness of protected areas has rocketed, as protected area agencies seek to justify and defend their budgets to reluctant governments, and private conservation organisations need to convince their members to keep donating. Initial questions about the effectiveness of biodiversity conservation soon developed into a broader analysis, as protected areas were also expected to deliver wider ecosystem services and social benefits. The emergence of different governance models, and different types of area-based conservation, further complicates the job for anyone trying to understand how protected area systems are functioning. The various ways of assessing management effectiveness and standards for management are described in this chapter, along with an attempt to provide an overview of the effectiveness of different types of area-based conservation.

The initial boom in designation of protected areas was based on an assumption of success, backed up by some spectacular examples. Iconic animals like the bison of North America, one-horned rhino of South Asia and some of the most magnificent ecosystems in southern and eastern Africa would have disappeared without the earliest national parks. There have therefore been some important successes in preventing total extinction in the wild.

Some species wholly or mainly dependent on protected areas

An increasing number of species are found only in protected areas, in extreme cases in just one protected area. The Alliance for Zero Extinction (AZE) lists species known from a single site; out of 853 AZE sites so far identified, 486 are in protected areas, but this is likely to be a small proportion of the total of species confined to protected areas. Above the Avon Gorge in Bristol, England, two endemic species of whitebeam (*Sorbus*, a broadleaved tree) grow in Leigh Woods, a national nature reserve that is their only known location on the planet (Rich and Houston, 2004). On the other side of the world, Wallangarra Whitegum (*Eucalyptus scoparia*, Maiden, 1905) is found only in Girraween National Park in Queensland Australia, growing in clefts in large granite outcrops. Girraween is on the AZE list, Leigh Woods is not.

Animals, being more mobile, are less likely to remain within the artificial borders of a protected area unless it is fenced, but a surprising number of even the largest and most iconic species are now mainly or completely reliant on protected areas for their survival. Most of the world's 4,000 or so wild tigers now survive in protected areas, many heavily guarded (Wolf and Ripple, 2017), although a gratifying increase in numbers has seen them straying farther afield in some countries. More surprising still, the lion, once ubiquitous in Africa, is now largely confined to national parks and hunting reserves with only ten areas (four in East Africa and six in Southern Africa) identified as offering a secure future for lions (Riggio et al., 2013). An increasing number of national parks in Africa are now actually fenced, including Kruger National Park in South Africa and Nairobi National Park in Kenya, meaning that lion populations inside no longer even have the opportunity for dispersal through the landscape (Pekor, 2019).

Other species are completely or partially dependent on protected areas. Recent research shows for instance that 84 per cent of African elephants (*Loxodonta* spp.) are in protected areas (Chase et al., 2016). The Cape mountain zebra (*Equus zebra zebra*) had an estimated population of 3,300 in 2016, almost all in protected areas (Birss et al., 2016). The black rhinoceros (*Diceros bicornis*) is confined to protected areas in South Africa, Namibia, Kenya, Zimbabwe and Tanzania, following a 96 per cent decline in population in the twentieth century (Moodley et al., 2017), while the Javan rhinoceros (*Rhinoceros sondaicus*) is entirely confined to Ujung Kulon National Park, Indonesia, with just fifty-eight to sixty-eight individuals surviving (Nardelli, 2016). Following extinction in the wild, the Arabian oryx (*Oryx leucoryx*) has been reintroduced into Saudi Arabia and a thousand animals now survive in Mahazat-as-Sayd protected area, which is completely fenced (Fisher, 2016). The wild Bactrian camel (*Camelus ferus*) is virtually confined to a few protected areas in Mongolia and China. And if plans in China go ahead, almost all the remaining habitat of the giant panda will be included in a large national park. There are many more examples, with research suggesting that protected areas are particularly effective in protecting mammals with ranges of under 10,000 km^2 (Pimm et al., 2018).

Some species, now no longer in danger, owe their continued existence to their conservation in protected areas through critical periods; the American bison (*Bison bison*) is one of these (Redford and Fearn, 2006). As species recover, the critical role that protected areas have played in the past is often forgotten.

Species losses in protected areas and continuing biodiversity decline

This doesn't mean that protection is perfect. As the protected area estate continued to expand more critical voices began to be heard, with research showing continued decline in mammal species within North American (Newmark, 1995) and African (Craigie et al., 2010) national parks. More generally, repeated analyses demonstrated that biodiversity continues to collapse globally at levels and speeds not seen for millions of years, if at any time in the planet's history (Ceballos et al., 2017). Coupled with the kinds of criticisms from human rights groups outlined in Chapter 11, this led to a more general discussion about whether conservationists were backing the wrong approach in assuming that protected areas would "save" wild plant and animal species.

Questions ranged from whether protected areas were working at all to whether their successes in conserving biodiversity were worth the human costs. Associated issues related to how protected areas were managed and governed. Some scientists claimed that only "strictly" protected areas were likely to be successful – places where human access was eliminated or tightly controlled. Others believe that cultural landscapes and seascapes, with people integrated into ecosystem management strategies, are both more socially equitable and likely to be more effective. Finally questions of governance emerged; is it better to have a centralised, government run protected area or one managed collectively by a community or indigenous peoples' group? Some of these questions are also deeply philosophical, driven and influenced in large part by the underlying beliefs of the person doing the comparison. We've talked to many people for whom the question of whether or not protected areas are effective is already settled in their own minds; some of these are convinced of their usefulness while others are equally convinced that they have been a costly mistake that should be rectified forthwith.

Changing commitments to area-based conservation

At a national level, governments play a critical role in determining the future of area-based conservation. Taking the broadest approach, attitudes to indigenous reserves, tribal lands and other common lands often determine the amount and speed of land conversion or access to marine resources. In many countries, local communities and particularly indigenous communities remain under extreme pressure and the phenomenon of land-grabbing has become widely recognised; some of this directly relates to natural and semi-natural ecosystems being expropriated and converted to uses such as soy feed for intensively reared livestock. (The concept of "green grabbing" also emerged recently to describe land taken from local communities for conservation purposes; we discuss this in the following chapter.)

Protected areas themselves are not immune to pressures from governments. Although they are set up on the assumption that they are permanent, the evidence suggests otherwise. The World Database on Protected Areas charts an annual increase but also a steady trickle of sites that are being de-designated for some reason or another. Sometimes governments change their minds, perhaps because valuable resources are discovered inside protected areas, or the ruling political party changes to one less sympathetic to conservation, or on other occasions protected areas managed by individuals or communities are abandoned as social attitudes change. Within many governments, ministries may disagree; it is far from atypical for a ministry of mining to issue prospecting rights through national parks without consulting or even informing the ministry responsible for environmental issues, which is usually politically weaker.

For the last few years Mike Mascia and colleagues have been charting the growth of Protected Area Downgrading, Downsizing and De-gazettement (PADDD) as conflicts about access to natural resources intensify (Mascia and Pailler, 2011). Protected area boundaries may be trimmed to make way for development; roads and railways are pushed through the middle and parks will also sometimes be opened for other forms of exploitation. There is often tension between tourism and conservation. Finally, in extreme but far from uncommon cases a protected area will simply be de-designated by a government. And similar changes in attitude and priorities may be found in protected areas run by indigenous peoples and local communities as well; some ICCAs have been abandoned when community composition changed and younger members with different priorities became more prominent in decision making. Since the start of the modern protected area movement, over half a million km^2 have been removed from protection and three times that area have had protection status reduced (Golden Kroner et al., 2019). For example, in 2015 PADDD proposals threatened $60,555$ km^2 in the Amazon (Pack et al., 2016); things are likely to be far worse today.

Even if protected areas are protected by law or by other effective means, they may not be implemented. The term "paper park" is used to describe protected areas that have been gazetted or otherwise established but never provided with the management plans, funds or staff to be set up as actual protected areas – they remain as lines on a map and may not even be recognised or known about by the people living inside or close to their borders. The term itself is controversial, with critics pointing out that designation is an extremely important step in the right direction, and provides a not inconsiderable level of protection, for example dissuading many companies from converting the area for mining and agriculture. But the concept is nonetheless important because it includes many protected areas that are simply not working, and a more modern definition of a paper park is: "a paper park is an area that has been legally or otherwise designated as a protected area, but has not implemented any processes to achieve the conservation of nature" (Jessica Stewart pers. comm.). We still don't have a clear picture of how many paper parks there are in the world – and any estimate would inevitably be imprecise given the amount of interpretation possible in the definition, but this remains a major problem.

A quick overview of pressures on protected areas

Pressures come in various forms as discussed earlier; they can be characterised as: "immediate pressures" such as poaching or encroachment; "underlying pressures" such as demographic change or the impacts of long-distance pollution; and what might be summed up as "future pressures", of which the increased impacts of climate change are the most well known. Immediate pressures can be further subdivided into external impacts on the protected area, internal pressures and resource exports from the protected areas. It is also useful to distinguish between threats that the manager of the protected area can have a reasonable chance of addressing themselves (noting that a "manager" can be anything from a government employee to a village council), through to issues that need national attention from a government, to those that require international responses. So, for instance managers can with luck reduce poaching pressure through careful enforcement policies but would usually need government help in addressing pollution, and global initiatives are needed to reduce the rate of climate change in any significant way.

External pressures

Many pressures have been summarised in Chapter 8; protected areas need to deal with all of these. The best-known today is probably climate change, which is presenting, and will continue to present, serious management challenges. Many coastal protected areas will simply disappear under rising sea levels and human population pressures will often prevent moving the reserve inland. (Conversely, some currently settled areas may become uninhabitable and "freed up" for conservation.) Ecosystems will change, and some species will find that their habitats or optimal living conditions have disappeared from the protected area that was established to conserve them. Increases in extreme events, such as storms, fires, floods and snowfall, will alter ecosystems even in places where they survive. Many glaciers will simply disappear – a number have already done so – with implications both for the associated flora and fauna and for downstream hydrology. Ocean acidification and warming will change and reduce marine biodiversity, potentially to a dramatic degree. Protected area managers may find themselves responsible for rapidly changing ecosystems, and there is an ongoing debate about what this will mean in practice, ranging from the need to shift the boundaries of reserves to keep up with changing conditions, move species to places where they will have a chance of surviving and bringing in management responses of varying levels to maintain systems such as managing fire or altering water chemistry (Hole et al., 2009; Gross et al., 2016).

Challenges do not only relate to practical management but also have a psychological element as conservation managers need to adjust to loss and change (Wyborn et al., 2016). Protected areas are predicated on the concept of holding onto a valuable ecosystem; if this is changing due to forces way beyond the manager's control it requires emotional resilience to manage as well as possible and embrace change when it occurs without losing heart. We'll look at these issues in more detail in the final chapter.

Managers often have little ability to reduce the impacts of long-range pollutants from air or water; seepage of fertilizers from farmland is one of the major causes of damage to the Great Barrier Reef in Australia, while reef mangers look on helplessly. A protected area is also no automatic defence against invasive species, although intact vegetation is often less susceptible to invasion by non-natives than degraded areas.

Other external pressures can come as a result of large-scale human migration, due for instance to social unrest or climate change, while conflict itself creates additional problems for protected areas, including increased poaching (Hanson et al., 2008). Conversely, some conflicts can maintain ecosystems in a relatively intact state if places become too dangerous to enter and peace can bring increased forest loss. Analysis of four case studies found an average increase of 68 per cent deforestation in the five years following cessation of conflict (Grima and Singh, 2019).

Transport links through or near protected areas can, unless sympathetically designed and well monitored, be a major cause of degradation by facilitating easy access into a protected area for poachers (Alamgir et al., 2017). Changes to hydrology, such as dams upstream or downstream of the protected area, can create changes by interrupting water flow or blocking fish migration (Finer and Jenkins, 2012).

Internal pressures leading to export pressures

Inside the borders, many protected areas are subject to pressures ranging from incursions and illegal settlement to various forms of plant and animal poaching for sale, bushmeat hunting and plant collection for food and medicines, firewood and fodder extraction and the accidental damage that accompanies these activities including particularly increased risk of fire (Carey et al., 2000). While many of these activities will be for local subsistence use or sale, others will be for export, for instance to the lucrative commercial bushmeat markets throughout the tropics, to criminal gangs dealing in ivory and skins, timber and wildlife products like coral and aquarium fish. Local use of resources from protected areas can be incorporated into management and given legal sanction, which can build support for protected areas amongst local communities, although the difficulties of doing this sustainably should not be underestimated. Elsewhere, the fact that wildlife becomes concentrated in protected areas due to its decline in the wider landscape or seascape means that these can become a magnet for people dealing in wildlife products. Twenty years ago, managers in Serengeti National Park, a reasonably well-managed protected area in Tanzania, were estimating to us that poachers were taking 200,000 animals a year and settlement has actually increased around the edges of the park due to the (illegal) bounty it contains (Veldhuis et al., 2019).

Even in situations where there are far less pressures – and it is misleading to assume that all protected areas are under siege – management input is usually needed for visitors, to monitor changes in biodiversity and the environment and often for restoration, changing habitats to attract particular species and so on. Most protected areas are not "wild" in that they are left completely alone but are managed for wildlife. In many parts of the world historical or prehistorical changes, such as the loss of top predators, means that the ecosystem remains out of balance and many protected areas require management inputs if they are to retain the species for which they were established. Some people argue that these efforts are misplaced and that we should let an ecosystem find its own new natural level – the rewilding movement is part of this. It remains highly controversial, but the concepts are gaining momentum in many parts of the world. The extent to which this really means walking away is still open to doubt; many of the designated wilderness areas in the United States continue to have some management input.

And we often struggle to understand what this input should be. The rate at which protected areas have been established has overtaken the rate at which professionals can be trained to undertake management, especially so as the amount expected of managers keeps increasing as protected areas are expected to fulfil an ever-larger range of functions. Many protected areas are facing a shortfall and suffering in consequence.

Assessing management effectiveness

The first step in addressing such a portfolio of issues is to find out what is happening, itself not an easy process. From a methodological point of view the question is complicated; it doesn't only matter if particular plant and animal populations are going up or down, or whether desired ecosystem services are being maintained, but also whether they are doing so *because* of the presence of the protected area. For example, if populations of a particular species have changed, are these trends different from those found outside the protected area? Data are scarce; even if accurate trend data are available inside a protected area there is unlikely to be the same level of detailed monitoring available in the surrounding areas. But if every case without a counterfactual (a comparison of the sort just described) is set aside then analysis will probably be based on a small and possibly atypical set of cases. And if clear trends in biodiversity are seen, is this because of management or happening despite of or independent from any management actions?

Twenty years ago, the IUCN World Commission on Protected Areas (WCPA) agreed a basic framework for assessing the management effectiveness of protected areas, which sought to put in the context of management any gains or losses in the values that the protected area was established to maintain (Hockings et al., 2006, 2nd edition). The WCPA framework is based on the premise that good management starts by establishing a vision, influenced by the "context" of existing status and pressures and progresses through "planning"; the strength of management systems is then examined with respect to allocation of resources ("inputs") and management actions ("processes"), and assessment finally considers achievement of objectives in terms of whether planned work has been completed satisfactorily ("outputs"), which hopefully results in the desired "outcomes".

Outcome evaluation determines whether management has achieved the objectives in a management plan, or national plans, and the aims of the IUCN management category. This involves long-term monitoring of the condition of the biological and cultural resources of the system or site, socio-economic aspects of use and the impacts of the management of the system/site on local communities. Such evaluation should consider whether the values of the site have been maintained and whether threats are effectively addressed. Outcome evaluation is the true test of management effectiveness, but the monitoring required is significant; many existing assessment systems fall down in this regard because of the time and expense involved.

Management effectiveness evaluation can give evidence of success or failure, for instance in managing pressures such as poaching or controlling weeds, and enable protected area managers to adapt management if necessary. A good assessment system will provide indicators of overall ecosystem health to identify effects of global pressures such as climate change and test hypotheses to cope with this. Including a range of stakeholders will enable indigenous people and other local communities to become actively involved in assessing management and facilitate greater ownership and support for the park. Stakeholder processes provide a mechanism for integrating scientific and traditional knowledge as well as the perceptions and experience of protected area managers into decision making. Assessments enable practitioners to share ideas and experiences and make cross-site comparisons with consistent data and give early warning of protected areas in danger. They also help to argue for funding and international support for these areas. Data on economic and other benefits of protected areas can be used to build public and political support.

The WCPA management effectiveness framework is a set of best practices around which evaluation systems can be constructed. In the years since it was agreed, several dozen assessment systems have been developed, most broadly following the framework in terms of subjects addressed. These have been implemented globally with records of over 27,500 assessments being carried out on Global Database on Protected Area Management Effectiveness (GD-PAME). They vary from quick evaluations designed to be completed in a day or so, to detailed monitoring systems suitable for sites prepared to dedicate extra time and resources to finesse their understanding of management effectiveness or facing particular problems that need analysis. Two that we have been involved in ourselves illustrate the range.

The Management Effectiveness Tracking Tool (METT) (Stolton et al., 2003; Stolton and Dudley, 2016) adopts a questionnaire approach, where users fill in a basic data sheet, identify primary pressures on the protected area and select answers from a series of multiple-choice questions relating to various aspects of management. The answers chosen each have a score which once summed give a total score for the area; for any question scoring less than the maximum, respondents need to say what they will do to make improvements and there is also space to list reasons why a particular answer was chosen. The end result is both a total score that can be compared in repeat assessments and, more importantly, a list of next steps in management. The METT can be and unfortunately sometimes is filled in quickly by individuals over the space of an hour or so, but these results are suspect. A better approach is to bring together several people, at least from the protected area but ideally including local stakeholders or people from neighbouring protected areas and reach consensus on each question. A good facilitator is essential, if only to keep up energy levels during what is a long process of negotiation and consensus building. Like any tool the METT can and should be modified to fit its purpose. While many countries use it as it is, others work to modify the questions or add extra sections to suit their particular situation.

For example, in Bhutan we worked with protected area staff through two workshops to adapt and modify the METT, and this was applied in a nationwide survey, which was coupled with a series of site visits to discuss findings directly with local stakeholders (Lham et al., 2019).

Assessing effectiveness in Bhutan

Tingtibi Range Station in Jigme Singye Wangchuck National Park covers 485 km^2 with eighty households. Here we sat down with eight representatives from two villages, plus seven local and headquarters staff. One village is a long-settled agricultural community while the other practised a hunter-gatherer lifestyle until settled cultivation was introduced by Japanese development workers in the 1980s. Overall the villagers felt that while there were advantages and disadvantages of being in a protected area, the advantages were greater. The main non-timber forest products collected are ferns, mushrooms and cane shoots; the latter requires a permit. There is no monitoring of NTFPs. Banning commercial logging (for *Pinus roxburgii*) was seen as a benefit because timber was still available to residents and more sustainable. Commercial fishing was once also banned, impacting four or five households; this rule has been relaxed. Attitudes to fishing are complex; some villagers disapprove because they receive no benefits and possibly for spiritual reasons. White bellied herons have been seen nearby, which is a further complication as twenty-seven of a global population of 160 known birds are in Bhutan.

Human wildlife conflict has increased dramatically, from deer, sambar and wild boar, but this is a national trend and not particularly connected with the protected area. Wild boar population has expanded dramatically, perhaps due to hunting and poisoning of wild dog. The villagers felt that the switch from shifting to settled cultivation increased problems and left some crops very difficult to grow: mustard is attractive to deer and potatoes to wild boar. Government compensation for losses is being replaced by a community-based compensation insurance scheme, but this had not yet been fully developed. If villagers report crop raiding, park officials visit to estimate damage. Control efforts include solar-powered electric fencing; 3 km had been erected and another 3 km needed. The situation is complicated by land abandonment, which increases habitat for problem animals and reduces the number of people available for guarding etc. If a problem animal comes within 200 metres of a field villagers are allowed to chase and shoot it (except for elephant), but not to set snares or poison. There have been discussions about bringing professional hunters, but this is unlikely to be successful in limiting overall numbers; furthermore, for religious reasons many people object to hunting and hunters are likely to be ostracised. Regarding climate change, villagers generally perceive that the weather is getting warmer and weather patterns and rainfall more irregular. This is impacting cropping.

Meetings between park staff and villagers take place about crop raiding, fishing and rules and regulations, mainly in the villages. Rangers focus on monitoring, foot patrols and collection of field data related to combating poaching. They have a network of paid informers. Priorities for staff and communities included more electric fencing, sorting out questions about fishing and potentially introducing biogas technology. Rangers felt that they needed better training on management and assessment of crop and livestock damage. **SS/ND**

The speed and simplicity of the METT means that it has now been applied several thousand time around the world and has also been subject to some major analyses (Geldmann et al., 2015). Whether all the assessments are truly accurate is less certain; the METT was designed as a quick tool to track progress in individual sites over time and is likely to be less correct in comparing between sites due to the subjective nature of the responses (Stolton et al., 2019).

The Enhancing Our Heritage Tool is an altogether more detailed assessment process, developed originally for natural World Heritage sites and field tested in ten protected areas ranging from Aldabra Atoll in the Western Indian Ocean to Keoladeo National Park, Rajasthan, India. The system consists of a "toolbox" with twelve different assessment systems ranging from ways of understanding the political context within which the area exists to protocols for monitoring biodiversity over time (Hockings et al., 2008). Sites that already have some monitoring in place would select those tools that fill gaps in the existing system, thus avoiding duplication of effort. Sites that had previously undertaken little monitoring, such as Aldabra Atoll in the Seychelles, Western Indian Ocean, took the whole toolbox and applied it as a single assessment system. The toolkit has since been adapted for use in cultural World Heritage sites as well and as we write is in the process of revision.

After conducting many such exercises, in small and large groups, sometimes using translators, in the field and in offices we have become convinced that the primary purpose of an assessment is to bring people together and provide space for them to talk. Although designers of different systems are often passionate about their relative advantages, the methodology is by a long way secondary to the assessment actually taking place. Given the day-to-day pressures on managers a one-day assessment process may be the only time in the year when they have time to think strategically about whether management is going in the right direction or if it needs adaptation. Assessments of management effectiveness are as much about giving managers and staff an opportunity to sit back and think about what they are doing as they are about collecting data or making reports.

Other forms of assessment

Both these and most other management effectiveness assessment systems concentrate mainly on getting management right, focusing on issues like planning, budget levels and budgetary control, staff training, availability and maintenance of equipment and so on. The more comprehensive systems also include biodiversity monitoring (one of the main weaknesses of systems like the METT is that they do not pay more than cursory attention to overall conservation outcomes) and options for stakeholder discussions. Particularly important here is the Spatial Monitoring and Reporting Tool, SMART, a combination of software, training materials and patrolling standards to help conservation managers monitor animals, identify threats such as poaching or disease and make patrols more effective. SMART relies on information routinely collected by rangers or others as they patrol, providing standardised data management that allows the creation of maps, analyses and reports to help managers prioritise limited financial or staffing resources, and track changes in activity over time.

More recently, human rights and social equity groups have been calling for assessment systems to be extended to include more detail on issues like social costs and benefits, treatment of local communities and quality of governance. The International Institute for Environment and Development developed the Social Assessment of Protected Areas (SAPA) methodology, a detailed, ten-week process of assessment through a series of workshops (Franks et al., 2018), and at the time of writing are working on a quicker assessment system equivalent to the METT. The IUCN Commission on Environmental, Economic and Social Policy worked with WCPA to develop an assessment system for governance (IUCN-TILCEPA, 2010). Social assessments are only really possible with a larger group of stakeholders and need local knowledge to ensure that all points of view and groupings are represented, that no-one is intimidated from making their point and that the right questions are formulated to avoid biasing results. Although not developed originally for protected areas, there are a plethora of approaches to help, for example in selecting participants and in how to interpret results.

Developing a monitoring plan

Over the years we have developed monitoring plans with a number of protected areas; one of the first was an ecological monitoring plan in Serengeti, Tanzania, working with Marc Hockings and Ephraim Mwamgomo; the following draws on this experience (noting of course that protected areas will normally also need to be monitoring an array of other issues such as impacts on local communities, tourism etc). First, the main aims of the protected area need to be identified; these should already be listed in the management plan. A list of measurable things that could describe these aims is needed. This will generally start off as a long list, which will cost too much time and money to monitor, so staff and others need to work to reduce this down to a suggested six to eight **"biodiversity targets"** which could be **species**: both "iconic" species (e.g. the lion) and species that say something about the state of biodiversity (e.g. are dependent on particular habitats or under threat); important habitats; or ecological processes such as migrations. Taken as a whole, these should describe the status of the protected area.

For each target, **indicators to measure trends in status and threats** need to be identified and agreed. This could be changes in population size or recruitment rate for species, or changes in habitat area or condition; there may be several indicators needed for each target and they could include pressures such as pollution levels or poaching levels, number of traps found etc. These form the core of the monitoring programme. A good monitoring system will also contain some **thresholds**, levels at which the manager knows it is important to act. For instance, in a protected area involved in lion conservation, some minor variations in population size will be expected but if population dips beyond a certain point then something may be going wrong and changes in management needed. Conversely, if a population of elephants grows too large then the chances of human-wildlife conflict may increase dramatically and some form of control may be needed, either translocating elephants to another site or a cull.

Hopefully established protected areas will already be monitoring some of this information, but there will almost certainly be data gaps; in new protected areas there will probably be nothing.

Monitoring only gives accurate data if it is done in the same way each time, even if the people responsible change. Information needs to be stored so that it can be accessed and used in assessments later. So, the next stage is to develop some **monitoring protocols** and a **data management system**, describing exactly how information is collected and stored. Finally, an initial assessment is carried out to set a **baseline** and then regularly assess status of and threats to each target.

Monitoring systems are expensive and time-consuming and tend to be one of the first things to be dropped if there is a financial squeeze. But this is a pity; research shows that monitoring is one of the most critical elements of good management, if used properly, because it gives the information needed to make adaptations to management as required. **SS**

Monitoring should draw on other research carried out in the protected area. However, this information often does not reach people on the ground; we have talked to many protected area managers of even very well-known national parks who have not seen one of the papers published about their reserves. As far as we can see few researchers send information back to the places where it was collected and few protected areas can afford to buy expensive online publications, even if they know they exist. For example, workshops with protected area managers in Madagascar found that they had little knowledge of species trends, despite the existence of research

(Pyhälä et al., 2019). We got so frustrated by this situation that we helped to put together a code of practice for researchers working in protected areas (Hockings et al., 2013), but this divide continues in many places.

Do protected and conserved areas work?

There have now been many tens of thousands of management effectiveness assessments carried out, although still only a small proportion of protected areas have been assessed. However, there is already a large enough sample for us to draw some conclusions.

There are a number of related questions. First of all, are protected areas in the right place? This is often left out of the effectiveness debate, but it is critical; if the world is simply protecting areas like deserts and icecaps that would probably remain undisturbed anyway the system is irrelevant. Then, are protected areas being managed effectively? Does this add up to successful conservation of biodiversity? Are certain protected area management approaches more effective than others; in particular is the so-called fortress conservation of strict protection more or less effective than more inclusive, community-based approaches? A closely related question might be: do different IUCN management categories or governance types of protected area perform differently? Then a separate line of research needs to focus on the human impacts of protected areas; have they increased or decreased human wellbeing, and how do resident and local communities view their protected areas? Finally, can we say anything comprehensive about ecosystem services and the wider societal and economic value of national or global protected area systems?

The short answer is that we can say something about some of these questions but many of them remain at best only vaguely answered. Furthermore, existing opinions are now so ingrained within many people that finding truly unbiased research is a struggle. This is not to say that people make things up or deliberately distort data. But the wording of research hypotheses, the way data are used and the assumptions driving the design of projects are all open to unconscious bias. We have an uneasy feeling that, particularly when addressing the trickier questions, many researchers simply reinforce their prejudices and find what they expected to find. Perhaps more significantly, because there is a mass of contradictory information floating around it is easy to quote selectively, thus reinforcing opinions within a particular interest group that "their" particular opinion is backed by research. Furthermore, many of the large data-focused research is carried out by people who have little or no experience of implementing assessment at the protected areas level, giving them little feeling for the potential drawbacks or inconsistencies of site-level data. Finally, there remains the problem as noted earlier that little research incorporates counterfactuals, that is to say what would have happened if a protected area or an intervention undertaken in a protected area had not happened (Ferraro, 2009). Given the obvious reality that we have our own prejudices, in the following section we try as far as possible to give an unbiased account of the state of effectiveness around the world.

Are protected areas in the right place?

There has been an enormous, perhaps disproportionate, effort put into identifying the most important sites for biodiversity conservation, and some of the main ones are reviewed in Chapter 5. However sophisticated the methodology or the algorithm used, the strength of analysis depends ultimately on the accuracy and volume of data available; better studied countries are more likely to feature simply because we know more. It follows that more peaceful countries, and those with good living conditions, tend to have more information and therefore feature higher on the list of important places to conserve than those facing serious issues of security or unrest. Our own rough analysis of prioritisation exercises, now a decade old, found that Costa Rica and Cameroon both featured far more frequently than surrounding countries like Nicaragua, El Salvador, Central Africa Republic and Congo. There is no obvious reason for this to be the case in ecological terms, but both are relatively politically stable, and both contain active research facilities.

Currently, researchers suggest that many protected areas are indeed not in the "best" places for conserving biodiversity, and probably not for conserving key ecosystem services either. As noted before, protection of "rocks and ice" is politically an awful lot easier than protecting lands and waters that contain important natural resources such as timber or could be converted to grazing or agriculture. So Northeast Greenland National Park is the largest protected area in the world, but currently an easy decision as nothing much else is going on there (it will be interesting to see if it remains a protected area when the ice melts). We have already discussed some of these issues in Chapter 8.

There is no doubt that we need protected areas that make a difference, nor that many governments perhaps understandably try to wriggle round international commitments by setting aside areas that no-one else wants. Such situations need to be challenged. But an over-emphasis on only the "best" risks undermining any conservation efforts that fall outside the hallowed ground identified by whatever particular prioritisation exercise is in fashion at the moment; if this was taken to an extreme most of Europe would be omitted for example, as until recently would areas like upland peatbog which we now realise are not unproductive ground but vital as carbon stores. We can say that many important species and ecosystems are left exposed to development and other pressures because they are not included within any form of area-based conservation, and this is undoubtedly a situation that needs to be addressed. But the way we do this and the associated messaging needs to be nuanced to avoid undermining necessary conservation that is taking place in other areas as well.

Are protected areas being effectively managed?

Although the UN World Conservation Monitoring Centre collects information on whether or not a protected area has carried out a management effectiveness assessment, it does not currently analyse the result. Almost 10 per cent of the world's protected areas have undergone a management effectiveness assessment, using one of sixty-nine different methodologies, but only 1 per cent of the global total have so far done so more than once and are thus able to track changes over time (UNEP-WCMC and IUCN, 2018).

There was a flurry of attempts about a decade ago to aggregate and analyse the results of multiple assessments. We did a quick survey of sites working with WWF (Dudley et al., 2007) and identified the most important aspects of management success as being: legal designation, clear management objectives, demarcated boundaries, and operational plan, a working budget and good monitoring. A study of threats facing ninety-two protected areas in twenty-two tropical countries concluded that most protected areas are successful in protecting ecosystems (Bruner et al., 2001). A global meta-study assessed management effectiveness evaluations from over 2,300 protected areas and found that 86 per cent met their own criteria for good management (Leverington et al., 2008), and other studies found lower levels of land clearing in protected areas (Nagendra, 2008), particularly in indigenous reserves (Nelson and Chomitz, 2009), with often strikingly higher forest cover (Joppa et al., 2008) and reduced carbon losses (Campbell et al., 2008). However, a survey of over 4,000 sites by Fiona Leverington and colleagues (2010) found that 40 per cent showed major deficiencies. Several studies since have tried to correlate management effectiveness "scores" with biodiversity outcomes, with mixed results (e.g. Coad et al., 2015; Geldmann et al., 2018).

There is still a lot to be learned, but, as best we can judge at present, protected areas generally work if they are properly resourced and managed, especially with respect to retaining natural vegetation. This can be seen by a look at Google Earth or its Earth Engine, where protected areas often stand out as islands of vegetation in an otherwise transformed landscape. Their success at conserving particular animal and plant species within the protected area is much more variable, and in some national parks and nature reserves are almost devoid of larger animals. Even in well-managed and adequately resourced protected areas, a sudden and determined influx of poachers can overwhelm rangers; the sky-high prices commanded by ivory, bones from large cats, skins and even wild meat over the last few years has resulted in a surge of wildlife losses even in places that have traditionally been safe (e.g., Poulsen et al., 2017). These issues are too large to be tackled by individual protected areas, but need concerted efforts by governments, police and the judiciary, working with local communities (IUCN SULi et al., 2015). More intractably, we are still only slowly starting to understand how protected areas might continue to function usefully under some aspects of climate change (Gross et al., 2016), or how to address invasive species within the boundaries of protected areas (Foxcroft et al., 2013).

Our understanding about effectiveness of community managed areas, ICCAs, indigenous reserves and other indigenous protected areas is much more partial. We have good evidence that indigenous reserves in some places, including the Amazon, are as effective, and often *more* effective, than state-run protected areas at maintaining forest cover in the face of development pressure (e.g. Nelson and

Chomitz, 2009). Much less is known about effectiveness in other areas, and indeed the whole sensitivity regarding indigenous rights has sometimes made people reluctant to investigate. Most existing reports are qualitative rather than quantitative. An overview of the small amount of existing research (Rao et al., 2016) found a mixed picture: some successes but also some failures. An interesting paper by Intu Boedhihartono (2017) compared the ways in which six different indigenous groups in Indonesia managed their traditional lands and found a wide variation, with roughly half following more-or-less sustainable management and the others over-exploiting and degrading their lands. Similarly, Neema Pathak's huge study of community-conserved areas in India (2009) found many inspirational examples but also reported that surrounding communities sometimes viewed these expressions of sustainable living with bemusement. None of this is to say that community conservation doesn't work – the anecdotal evidence and our own experience shows clearly that it is an immensely powerful force, but that we still have little in the way of concrete evidence about when it does and does not.

In a slightly more specialised analysis, a group of us looked at evidence for the effectiveness of sacred natural sites in terms of delivering conservation (Dudley et al., 2009); examining hundreds of published analyses we found that many did indeed conserve biodiversity and could play a critical role in conservation strategies. The fact that scientists have worked with local communities to monitor biodiversity in SNS means that their value for conservation is much more generally recognised; a similar type of assessment is urgently needed for other forms of community-based management.

Ecosystem services

Studies of ecosystem services fall into three main types: economic assessments of all values including theoretical or as yet unrealised values; global studies of ecosystem service values; and economic assessments of money generated in particular places (sometimes including money saved, for instance for flood control or avalanche stabilisation). Economists like Robert Costanza and Ida Kubiszewski have led the way in total economic valuations of nature (e.g. Costanza et al., 2017), which have been very influential in drawing attention to the values that we are squandering. The TEEB process (The Economics of Ecosystems and Biodiversity) took the debate closer to governments and other policy makers and included an explicit focus on the values of protected areas (Kettunen, 2011). Tools like the Natural Capital project (NatCap) from Stanford University and partners, Co$ting Nature from Kings College in London and the Toolkit for Ecosystem Services Sites-based Assessment (TESSA) from BirdLife and partners all provide a series of packages and methodologies to work out the value of natural capital in particular places (see Neugarten et al., 2018 for an overview of many tools).

The value of a tiger

We collaborated with Khalid Pasha, Mike Balzer and other colleagues at WWF to work out the value of ecosystem services in tiger reserves, which include water, carbon storage, disaster risk reduction, medicinal plants, fodder, fish stocks and biological control (WWF, 2017). The role of tigers in supporting a tourism industry is also very important across the tiger range in India and elsewhere (Carter and Allendorf, 2016), although benefits are unevenly distributed (Reddy and Yosef, 2016). For instance, in 2014–2015, 1.47 million people visited Indian tiger reserves, 32 per cent of visitors to wildlife reserves in India (Karanth et al., 2017). Ranthambore National Park was India's first designated tiger reserve, in an area of dry forest in Rajasthan. Tigers are a major tourism attraction, initially mainly for foreign tourists but switching increasingly to domestic tourism; by 2007–2008, 80 per cent of nature-based tourism was domestic, although Ranthambore attracted proportionally more foreigners than other sites (Karanth and de Fries, 2011). Conversations that we had when we were there in 2012 suggested that many urban-based Indians now desperately want to see a wild tiger. The surrounding area supports 3,000 tourist beds and tourism revenues of 36.74 million rupees (US$500,000 per year, Verma et al., 2017), and there are also guiding activities and other benefits. Total revenue from Ranthambore was estimated at US$3,163,753 in 2011 (Karanth and de Fries, 2011), and tourism continues to increase with 374,134 visitors in 2015 (Karanth et al., 2017). An economic analysis of Indian tiger reserves identified other important benefits, including annual water services at 115 million rupees (US$1.6 million) and carbon storage at 69 million rupees (US$936,000) (Verma et al., 2015); however many of these benefits do not accrue to the protected area or local community at present. **SS**

In our own work on valuation, we have tended to start from the other end, asking local communities what they value from a protected area. The Protected Area Benefits Assessment Tool (PA–BAT) is the simplest of all the methods described so far: a set of questions about a range of potential benefits. The PA–BAT uses a standardised, workshop-based approach to gather information about a protected area's ecosystem services from a diverse group of stakeholders, reaching decisions by consensus. Around twenty-four questions cover benefits related to a wide range of ecosystem services, although users are encouraged to modify as needed. Participants explore current and potential importance (economic and non-economic) and find out who benefits and where the services are found in the protected area (Stolton and Dudley, 2009). Prior to the workshop, managers and others identify which benefits need to be assessed and approach different stakeholders (fifteen to thirty people) to take part. Key benefits are summarised in the workshop by use of a computer and projector, or writing up on sheets of paper, or maps, or even through quickly drawn cartoons. Facilitators are open to hearing about other benefits, and oral testimony is recorded; these stories are often the most useful part of the exercise. Assessments can be recorded and projected at the same time to create a highly transparent decision-making process, or summarised in cartoons and words by an artist, or benefits located directly onto maps. Local knowledge often – in fact in our experience usually – brings information previously unknown to protected area staff, itself illustrating the limit to communication that often exists between parks and communities. But local communities may also miss certain aspects, particularly global aspects, of benefits and some parallel literature research is often beneficial to complete an assessment, with any suggested changes checked with those who took part.

Using the PA BAT in Montenegro

Lake Skadar, the largest lake in southern Europe, lies on the border of Montenegro and Albania and is named after the Albanian city of Shkodër. Roughly two thirds is within Montenegro, and the rest in Albania; both sides are protected areas. It is a key biodiversity area with seven endemic species of fish and rich birdlife, including some of the last pelicans in Europe. Surrounded by spectacular karst mountains, the road along the southern side is narrow, beautiful and infrequently travelled. We went to the park headquarters in Montenegro with Kasandra Ivanic and Andrea Štefan from WWF in Zagreb to run a workshop to identify local perceptions of benefits. A mixed crowd of fishermen, hoteliers, local businesspeople and park staff took part. The following summarises the results of discussions along with some subsequent research on economic values.

Management involves extensive cooperation with Albania (Vujović et al., 2018). Many ecosystem services are recognised although as far as the stakeholders present knew, not all have been quantified in economic terms. The lake provides much of the water supply for coastal Montenegro with a new aqueduct bringing water to the coast at a rate of 1,500 l/s (Selulić et al., 2017). Honey production in the region involves around 7,500 hives and produces approximately 80 tons of honey a year, calculated at €0.83 million (US$0.96 million) a year (UNDP and GEF, 2011). Fish production is very high, running at 80 kg/ha/year. Fishing supports around 400 people. Some 300 of these catch bleak (*Alburnus* spp), operating for nintey-five days a year with a total annual catch of 456 tons, while another hundred catch carp (*Cyprinus carpio*), operating for 190 days a year and catching 95 tons. At a market price of €3/kg for bleak and €5/kg for carp, this works out at €1.8 million (US$2.1 million) a year. Some of the catch is used for the manufacture of value-added products, mainly canned or smoked fish with a production value of €1.4 million (US$1.6 million) a year (UNDP and GEF, 2011). An additional 300 families depend indirectly on the fishing catch (Mrdak, 2009). Some sixty families make all or most of their income through cruise tourism on the lake (Selulić et al., 2017), and there are many hotels and restaurants close to and within the national park; these benefits have not been quantified. Focusing on just fish production and honey, the national park produces over €4 million (US$4.6 million) a year. These values omit ecotourism, and local values are likely dwarfed by the total value of water to the country. **SS/ND**

There have now been many assessments of the economic benefits of protected areas; we have been collecting examples for the World Commission on Protected Areas and the Convention on Biological Diversity. A judicious economic analysis can help fend off funding cuts if the protected area system shows itself to be of net benefit for the community; this happened in Finland a few years ago (Kajala, 2012). Studies of the carbon storage and sequestration in protected areas have also become more common, in large part because of the increased funding opportunities from the carbon market. But so far there has been no global or even regional study of the benefits of ecosystem services from protected areas in general; this remains a lack that should be addressed.

Meanwhile, protected area managers, communities, private trusts and others are gradually building their expertise and understanding about lessons learned from management: what works, what doesn't work and what happened when things do go wrong.

What is important?

Management effectiveness assessments tell us when things are going wrong, but they do not necessarily provide detailed information about how to address any problems identified, nor do they set standards against which to measure results and document change. The CBD identified the need for such standards as long ago as 2004 in its Programme of Work on Protected Areas (CBD, 2004).

Standards can provide clear guidance on minimum requirements for effective management, specify monitoring protocols and include other considerations such as development of social safeguards. There is good evidence that standards improve adoption of best practices, lesson learning and the quality of results (Polasky, 2015). Standards also inform donors about whether their investments are being used effectively (Ferraro and Pattanayak, 2006; McCarthy et al., 2012), for both institutional and private donors (Bennett et al., 2015).

We started our professional life together working on standards and our very first joint publication was on *Guidelines for Conservation* for organic farmers and growers in the 1990s, looking at how organic certification could take greater account of nature conservation issues (Soil Association, 1990). In 2005, WWF asked us to look at what was known about management effectiveness at that time and suggest some of the most important elements of management; a slightly updated version of our analysis is given in Table 10.1 (and is not listed in any particular order of importance).

More recently we took part in many of the initial discussions in the development of IUCN's ambitious Green List of Protected and Conserved areas, which aims to promote and recognise effective management and good governance around the world. The Green List has set minimum management standards and establishes national committees to refine these for conditions in a country. Independent assessors then judge management against the Green List standards and, if the standards are met, a Green List certificate is awarded.

TABLE 10.1 Key requirements for effective protected area management

Minimum requirements	Details
Legal designation	Ensuring that protected areas have been gazetted by law. In many countries this is running behind the rate of protected area establishment, weakening the position of protected area agencies in the case of e.g. attempts to mine or drive roads through protected areas.
PA boundary demarcation	If protected areas have no legally defined boundaries it is difficult to stop encroachment or exploitation of resources at the edge and the total area can shrink. Demarcation is a key step in establishing the protected area and ensuring that it is properly managed
Management objectives	Many protected areas were set up with only vague objectives, thus hampering management. A number of tools now exist for working with managers and other stakeholders in setting both long- and short-term targets for specific protected areas.
Operational plan	Ideally protected areas should have a management plan, but this takes time and money. In the absence of a full plan, development of an operational plan can be an interim step to ensure that work follows an agreed pathway.
Operational budget	Most protected areas need a proper budget, and a surprising number do not have one or have only minimal understanding of budgeting and accounting. Basic business planning advice can help.
Monitoring plan	Research into integrated conservation and development projects found that a good M&E system was often *the* key to success or failure because it allowed critical adaptive management to be used.
Indigenous people and local people	Failure to address the wants and needs of local communities, including particularly indigenous peoples, quickly leads to resentment, problems and pressure on the protected area. Many protected areas fail to engage thoroughly or in a participatory manner with local communities.
Illegal use of resources	Of critical and growing importance. Illegal use comes in two forms: daily use by local people, which needs to be addressed through local community approaches and agreements; and high-end, professionalised poaching of valuable species of plants and animals, with the ivory trade being a classical example.

We have also worked from its initial conception on the Conservation Assured | Tiger Standards, which looks more narrowly at whether protected areas within the tiger range are effective in protecting tigers. The concept is now being extended to other species, such as the jaguar.

Conservation Assured | Tiger Standards

When a species in the wild falls to not much more than 3,000 individuals you feel especially lucky to have a "close encounter". Tigers (*Panthera tigris*) have been virtually driven to extinction by poaching (Stoner and Pervushina, 2013), habitat loss and fragmentation (Walston et al., 2010) and loss of prey species (Damania et al., 2008). The wild population has fallen by over 95 per cent since 1900, from around 100,000 to a historical low of 3,200 individuals (Harihar et al., 2018), and are found in around 7 per cent of their historic range (Walston et al., 2010). Watching a male tiger padding gently through the forest, as we did in Ranthambore National Park some years ago, is an increasingly rare treat.

The story of saving wild tigers is an exemplar of governmental coop-eration. In 2010 tiger range countries committed to doubling the global wild tiger population by 2022, an ambitious goal and the motivation behind much conservation action. A key strategy identified was to secure stable tiger populations in effectively managed protected areas (Walston et al., 2010). Recognition that many protected areas in the tiger range are not well man-aged (Conservation Assured, 2018a) means that measurable steps to improve management effectiveness are a critical component of tiger conservation strategies. Although some tiger reserves already carry out management effectiveness assessments, for instance in India the MEETR – Management Effective Evaluation for Tiger Reserves (Mathur et al., 2014), management effectiveness tools offer less in terms of measuring management against best practice, nor do they provide explicit guidance on how better to manage for tiger recovery.

We have been involved in the Conservation Assured | Tiger Standards (CA|TS) from the first concept note to its regional roll-out. The aim of CA|TS is to provide expert-driven species-specific conservation standards to both drive and measure progress towards improved tiger conservation (Conservation Assured, 2018b). It is an addition to, rather than a replacement of, management effectiveness assessment. CA|TS supplies minimum stand-ards for good tiger conservation and an independent verification process to ensure sites are meeting these standards. CA|TS is run by National Commit-tees in tiger range countries and overseen by an International Committee to ensure equivalence across the range. At the time of writing, CA|TS is being implemented across some seventy sites in seven countries (Bhutan, Bangla-desh, China, India, Malaysia, Nepal and Russia), where National Committees have been established. Six sites have already attained CA|TS Approved status.

And there has been some good news; tiger numbers in India are stable and show some increase (Jhala et al., 2019), and similar successes are being reported from Nepal, Russia and Bhutan. But the challenge of finding space for wide ranging predators such as tigers in an ever more crowded planet is daunting. My other very close encounter with a mature male tiger was not so happy. One eye glared at me through a tiny gap in a wooden enclosure and then a roar reverberated through my body – a feeling like standing by a speaker at a rock concert. This tiger was angry and vicious; he was terrorising a local community in a protected area that had worked hard to ensure community support for its conservation activities. For this tiger who wanted to roam far and wide there was simply not enough space. **SS**

Financing protected areas

Finance is the great bugbear of protected areas. Governments set them up but then often assume that they can more-or-less run themselves; conservation is an early victim of many cutbacks. A decade ago, when we analysed public funding for protected areas in fifty countries (Mansourian and Dudley, 2008) we found funding to be generally on the decrease, despite commitments from donor countries and despite there being more protected areas than ever before. Funding decisions by governments generally ignored the value that biodiversity and ecosystem services that area-based conservation brings, nor are funds necessarily matched very closely to need. The situation has not improved since. In many of the protected areas we work in, rangers lack the most basic equipment, impacting their personal safety as well as their ability to achieve their work; for instance, often struggling to find money to buy petrol for their vehicles, if indeed they have vehicles at all. WWF recorded 871 ranger deaths in the line of duty between 2009 and 2018 (Belecky et al., 2018). Lack of funds also impacts maintenance; equipment bought with a grant often has a short lifespan because there are insufficient resources for upkeep, or knowledge among staff about how to upkeep. Lack of investment also impacts staff; there is often little training for rangers and poor wages: under-skilled and under-valued people do not help overall effectiveness. Even in some of the countries where protected areas are a major, or the major, source of revenue, management is often squeezed to a minimum. Annual funding shortfall for protected areas containing lions for instance was estimated at more than US$1 billion (Lindsey et al., 2018).

In part this is a matter of societal choice. Estimates for total budgets needed for protected areas sometimes sound large but are dwarfed by what governments instinctively spend on defence, major infrastructure projects and the like. Nor is the situation remedied in any simplistic way. Grants that provide major funding to individual protected areas to build capacity and set them on the road to success are fine in principle but can backfire in countries with high levels of corruption and poor rule of law; we have seen good managers abruptly pushed out of their posts and replaced by a relative of the minister's and much of the money squandered or siphoned away. In these circumstances decent staff members quickly drift away and the situation can be worse rather than better.

Private funding is a possibility but is challenging. Funding projects, like the Amazon Regional Protected Area programme, ARPA, brought large sums of private money into Amazon protected areas on the understanding that once these were exhausted the Brazilian government would take over but a current environmental backlash in Brazil must bring this into doubt. Groups like the Wildlife Conservation Society find that if they start financing a protected area the government is likely to expect this to continue indefinitely. There is also the problem that many people running protected areas are more skilled in wildlife and natural resource management than in budgeting, so that the money available may not always be used as effectively as it should be. Much budgeting is still in the realm of fantasy. We have seen too many management plans, more often than not completed by external consultants, that are wish lists of interventions and projects that meet all the current thinking around "best practice" protected area management, which would in theory be wonderful if a major pot of money materialises but are in practice unrealisable, rather than a realistic set of plans for what can be done now with the money available, or with a realistic increase.

Protected areas are rightly coming under increasing pressure to raise some of their own funds through various fees structures, exploitation of ecosystem services that are compatible with management and private donations. We support this, although note that it will only ever be a partial solution for most countries. Areas that are important for key species or ecosystems may have little attraction for visitors and few ecosystem services but still be crucial from the perspective of biodiversity conservation. In some cases, protected area agencies may be able to cross-fund from other more popular reserves. In the Seychelles, most of the revenue for managing the remote Aldabra Atoll World Heritage site comes from the entrance fee to and sales from the much more accessible Vallée de Mai, the country's other natural World Heritage site. Gorilla watching in Bwindi Impenetrable Forest National Park supports most of the rest of the Uganda national park system, but this is also a high-risk strategy. Changes in fashion amongst tourists or something unforeseen like a terrorist attack can radically reduce the funding available; Uganda underwent a funding crisis in the late 1990s when some tourists were kidnapped and killed, and tourism collapsed for several years.

Nonetheless, many governments are actively looking to build viable financial models, a few of which are outlined in the following.

User fees

Many protected areas – globally most national parks for instance – charge people to visit; the same is true for many privately protected areas if they are run by non-profit trusts. In countries where ecotourism is popular such fees can provide ample funds for conservation, while in others they are merely a contribution. Governments differ markedly in the extent to which fees are returned to sites for management. Protected area agencies have been innovative in finding alternative ways of extracting money from visitors (Spergel, 2001), including charges for specific activities such as wildlife viewing, guided hikes and diving fees; receiving a small percentage of profits from beneficiary businesses such as tour companies, hotels and cruise ships; various fees for ecosystem services such as carbon and water; and more controversially, fees from hunting concessions. Elsewhere, innovative approaches such as lottery revenues, wildlife license plates for cars, revenue from advertisements that feature animals and fines for poaching or pollution all contribute on occasion.

Hunting deserves special mention. In some places, particularly in Africa, private game reserves are managed for hunting, maintaining wildlife populations and charging hunters a hefty fee to shoot under controlled conditions. There is an active debate at the moment about whether or not hunting concessions are critical to conservation in Africa for instance (Gaskill, 2019). Hunting concessions cover around 100 million hectares in Africa (UICN/PACO, 2009), providing incentives for retaining land for wildlife while also protecting wider ecosystem services. In some countries, e.g. Zimbabwe, catering to a small number of high-paying foreign tourists was seen as a necessity when political instability reduced the number of photo-tourists (Funston et al., 2013). Hunting fees and revenues are substantial for those directly involved; lion hunts attract the highest mean prices (between US$24,000 and US$71,000) of all trophy species and generating significant revenue for wildlife authorities (Lindsey et al., 2012), with estimates for eight countries (South Africa, Namibia, Zimbabwe, Botswana, Ethiopia, Mozambique, Tanzania and Zambia) of around US$132 million (Murray, 2017) a year. In West Africa for instance, just over 40 per cent of the vast W-Arly Pendjari ecosystem of Benin, Burkina Faso and Niger is leased as hunting areas to private operators who take over management responsibilities for the areas. Around 75 per cent of the meat harvested is provided to neighbouring communities, between 30–50 per cent of revenue is fed back to local associations and the concessions are major employers (Bouché et al., 2016). But the practice is deeply divisive. Lions are often hunted in areas adjacent to a fully protected source population (Creel et al., 2016). The killing of an iconic lion, in the Hwange National Park in Matabeleland North, Zimbabwe, which had wandered outside the boundary caused global backlash (and death threats against the hunter) (Lindsey et al., 2016). Proponents argue, with some justification, that habitats and wildlife in hunting concessions are better protected and populations more stable than in many poorly resourced protected areas, and that the offtake is a price worth paying. Many conservationists, and many members of the public, baulk at setting up "protected areas" and then charging super-rich people to kill endangered animals. Several countries ban lion trophy imports, and some airlines will not transport trophies. Revenues from hunting may decline as pressures against hunting grow, leaving a major gap in funding (Bouché et al., 2016).

Payment for ecosystem services

Many of the ecosystem services described in Chapter 9 can also raise funds, either directly because people are permitted to collect products that they sell or through some form of Payment for Ecosystem Service (PES) scheme. PES schemes match users of ecosystem services with those carrying the costs of maintaining that service and arrange a fee. PES includes REDD+ schemes conserving carbon (see the following section), biodiversity conservation and water schemes, where companies or municipalities pay people upstream to keep the watershed intact. Cities such as Mexico City (Caro-Borrero et al., 2015), New York (Environmental Protection Agency, 1999), Beijing (Scherr et al., 2006) and Quito (Pagiola et al., 2002) use PES as a way of securing ecosystem services. Other PES options include various agri-environmental schemes, payment for biodiversity conservation and bioprospecting, where agriculture or pharmaceutical companies pay to access wild species for research purposes (Dunn, 2010).

Lessons learned from around the world can be instructive. PES schemes have been running for twenty years in Costa Rica, where it is calculated that they have conserved almost a million hectares of forest. Benefits to communities are both direct, in terms of payments, and indirect through healthier ecosystems. PES schemes are increasingly targeted at areas most needing conservation. To be successful they need clear beneficiaries and payees with schemes ideally targeted at specific groups for the greatest social impact. A simple method for validation of environmental impact is important. Cost effectiveness can be increased by linking to opportunity costs, ensuring a fairer distribution of available funds. Continuous monitoring is needed (Porras et al., 2013).

Financing mechanisms related to carbon emissions

Protected areas have great potential to help reduce global greenhouse gas emissions and to benefit from the reduced emissions from deforestation and degradation (REDD) financial mechanisms being developed within formal and informal processes linked to the UN Framework Convention on Climate Change. In particular, the REDD+ mechanism is aimed at maintaining standing forests and providing social benefits. A few years ago, there were hopes that REDD would solve the problems of protected area financing worldwide. But although there have been some interesting schemes that have worked, and are working, it has not proved to be a universal panacea.

REDD mechanisms aim to reduce emissions by providing compensation for "avoided deforestation", preservation of standing carbon stocks, and carbon sequestration through reforestation and afforestation. REDD has the potential to address several critical issues within a single mechanism: mitigation of global warming, reduced land degradation, better biodiversity conservation and increased human wellbeing and poverty alleviation. Institutions such as the World Bank are investing in REDD+ projects that will require capacity building and continuous, predictable and long-term funding.

Using carbon funds to support protected areas in the Russian Far East

Three protected areas in the Russian Far East have used innovative funding mechanisms to support local communities while simultaneously conserving Korean pine forests and populations of Siberian tigers. A presidential order banned logging of Korean pine, thus enhancing traditional pine nut harvesting, which has been further enhanced by developing facilities for processing, storage and marketing (BMU, undated). Pine nuts are economically important for local indigenous communities and are also the main food of the wild boar, which is itself the main prey of tiger. Existing pine nut harvesting zones have been enhanced through combating illegal logging and increasing protection. Forest protection prevents emissions of an estimated 130,000 tonnes of CO_2 annually. Income from carbon storage is generated through carbon credits under the Verified Carbon Standard. Water services are derived from the headwaters of the Samarga River. The protected areas have also secured traditional resources uses, which are co-managed by the government and indigenous Udege and Nanai people. Local communities are permitted to collect agreed amounts of other valuable non-timber forest products, including ginseng, various ferns, juice, seeds and frozen berries of magnolia and bog bilberries (UNESCO, 2017). Establishment of Bikin National Park has been described by indigenous leaders as a major step in securing indigenous rights (Sulyandziga, 2017). In 2014, the pine nut harvest earned villagers around US$60 million (WWF, 2017). Verified Carbon Standard funds to compensate for the Sochi Olympics' emissions earned the community over US$550,000, which paid for various community projects (BMU, undated). **SS**

However, there are some problems. Much destructive forest loss and degradation is illegal, and there is no reason to suppose that countries undergoing rapid deforestation have strong enough governance to address this problem (Saunders and Nussbaum, 2008). REDD investments in areas that are later deforested illegally are wasted. Some analysts fear that badly managed REDD projects will increase pressure on poor communities in terms of security of land tenure and access to resources by making areas they depend on for their livelihoods off limits (Mehta and Kill, 2007). Depending on how the details of the mechanism are worked out, these problems could encourage investors to put their REDD money into the safest options, which are usually not those forests facing the most acute threats, so that while it will provide useful resources it will not necessarily reduce deforestation as much as was hoped. Some activist groups and indigenous peoples' organisations have stated opposition to REDD on the basis that it will rely on sacrifices made by the poorest people rather than cut energy and fossil fuel consumption by the world's rich (Word Rainforest Movement and Grain, 2014).

That said, there are some encouraging uses of REDD+ projects emerging. We have been impressed talking and listening to the brains behind the Bio Carbon Partners in Africa, which deserves to be upscaled. Using a project approach based on partnership with communities and government, the project is creating important wildlife corridors across more than a million hectares of Zambian forest. A key issue with REDD+ projects is that they are performance-based, incidentally making projects outside of protected areas candidates for OECM status. They are audited and "verified" on a regular basis to ensure that forest protection is taking place.

Conclusions

The world has moved a long way in the last twenty years, in understanding both the threats to area-based conservation and how these might be managed. Assessment systems are becoming increasingly sophisticated, although there are still gaps in both methodologies and in coverage. In particular, it is notable that almost all the discussion in this chapter has been about protected areas and state-managed sites; we know much less about the effectiveness of other governance types and almost nothing about newer designations such as OECMs. Standards are being set and applied; it is still far too early to be sure how widely such processes will be adopted although the signs are encouraging. Eventually management effectiveness relies heavily on financing, and new financing options are being explored and developed. But gains remain fragile and subject to change; if area-based conservation is extended in the way that many people believe is essential then an order of magnitude more work will be needed in the future.

The last Javan rhino in mainland Asia

Cat Tien National Park (CTNP) lies 150 km away from Ho Chi Minh City in Viet Nam, through southern border towns with their largely Christian populations; many houses have life-size statues of Jesus on the roof. The protected area covers 720 km² along the Dong Nai River, protecting one of the largest remaining lowland rainforests in southern Viet Nam, a mixture of primary and secondary forest and seasonally flooded grasslands.

The Javan rhinoceros (*Rhinoceros sondaicus annamiticus*) was rediscovered by George Schaller in Viet Nam in 1989 (Schaller et al., 1990), having previously been presumed extinct in mainland Southeast Asia. Its viability was always in doubt, but there was pressure to conserve this unique sub-species. The rediscovered population was close to CTNP and following recommendations from WWF and the IUCN Species Survival Commission (Foose and van Strien, 1997) CTNP was expanded to include the main base of the rhinoceros. While total population was never confirmed, villagers saw a female give birth in the 1990s. But despite investment of a lot of time and money, the population continued to decline and the last individual died, probably after being shot by a poacher, in 2010 (Anon., 2010).

CTNP is both a Ramsar wetland of international importance and a UNESCO biosphere reserve. It contains many rare and endemic plant and animal species including nationally or globally threatened species. It is relatively well staffed, with an established management board, headquarters, visitor accommodation and many foreign visitors. WWF supported a full-time technical advisor for seven years along with multiple research projects. Two WWF researchers, Sarah Brooke and Simon Mahood, were investigating Rhino population size using a tracker dog to find dung samples when the last animal was killed. Despite these efforts, the park did not successfully protect its most important species. What went wrong? WWF asked me to revisit Cat Tien and compile an analysis (Dudley and Stolton, 2011).

Three underlying factors put the rhino under pressure. Parts of CTNP were heavily sprayed with dioxin herbicide ("Agent Orange") during the American war, with long-term ecological consequences (Vo, 2005) including loss of many trees and heavy growth of rattan and bamboo, reducing habitat for rhinos. Human population pressure increased dramatically, in part because of southward migration after the war. Many former soldiers were resettled near Cat Tien, and there was spontaneous movement of Catholics and several ethnic minority groups to avoid discrimination. Existing villagers found themselves squeezed by incomers. And at the time a "freshly cut" rhino horn was worth US$25,000–$40,000/kg in Viet Nam (Save the Rhino, 2010). 2010 was a particularly bad year for rhino poaching, with an estimated 232 taken in Africa and Viet Nam being a major destination (Anon., 2010). Vietnamese officials had been deeply implicated in trade. In CTNP patrols found 450 snares over a one-year period, around 10 per cent large enough potentially to trap a rhino (Anon., 2010). Poachers also use guns, and villagers in nearby Vinh Cuu reserve used poison against elephants (killing approximately half the remaining population during 2010).

National and local factors complicated the situation. Viet Nam has an admitted problem with corruption with respect to control of poaching and conversion of land. From 1996–2007 some 635 tons of wildlife and 181,670 individuals (animals) were confiscated (Nguyen et al., 2007), recognised to be a tiny proportion of the total. Patrolling, which was a major focus of WWF's capacity-building exercises, is reported to have declined sharply whenever WWF was not actively engaged in conservation.

Political influence is divided between three provinces that do not always coordinate well, nor agree on all issues of management. The government decided that the only way to safeguard the rhinos was to resettle 209 households and purchase land from another 140 households; compensation was secured but nothing happened, due largely to internal disagreements about strategy.

There was also increased activity by human communities including agriculture in the reserve and the buffer zone, which is now largely converted (Anderson et al., 2009). The rhinos were confined to around 5,100 ha of sub-optimal habitat. Many of the people I interviewed felt that this was already too small an area to support a viable rhino population and even this was subject to increasing disturbance, with important salt licks and wetlands destroyed. Members of the protected area Management Board felt isolated and under pressure to accept activities that they knew were detrimental to wildlife. While WWF and other organisations were pushing a conservation agenda, other development organisations were promoting development that proved to be counter-conservation, including particularly agricultural development. Different departments of the World Bank were simultaneously supporting both conservation and development projects in the region with apparently little coordination.

Maybe the population was doomed anyway. But other rare populations have come back from similarly small numbers; indeed, other rhino populations in Africa have been rebuilt from a handful of individuals. But the combination of multiple pressures, political infighting and lack of focus eventually doomed the rhinoceros. I came to the conclusion that when it came down to it, everyone simply gave up. **ND**

References

Alamgir, M., Campbell, M.J., Sloan, S., Goosem, M., Clements, G.R., et al. (2017) 'Economic, socio-political and environmental risks of road development in the tropics'. *Current Biology Review*, vol. 27, pp R1130–R1140

Anderson, Z., Hirsch, P., O'Connor, S. and Sunderland, T. (2009) *Integrating Problem Formulation in Tools, Methods and Approaches for Articulating Trade-Offs in Conservation and Development: The Case of Cat Tien National Park, Vietnam*. Workshop by CIFOR, MacArthur Foundation and ACSC

Anon. (2010) 'Patrolling project comes to an unsatisfactory end in Cat Tien national park'. *The Rhino Print*, p 4

Belecky, M., Singh, R. and Moreto, W. (2018) *Life on the Frontline 2018: A Global Survey of the Working Conditions of Rangers*. WWF, Singapore

Bennett, J.R., Maloney, R. and Possingham, H.P. (2015) 'Biodiversity gains from efficient use of private sponsorship for flagship species conservation'. *Proceedings of the Royal Society B: Biological Sciences*, vol. 282, pp 1–7. Doi:10.1098/rspb.2014.2693

Birss, C., Cowell, C., Hayward, N., Peinke, D., Hrabar, H.H., et al. (2016) *Biodiversity Management Plan for the Cape Mountain Zebra in South Africa*. Version 1.0. Jointly Developed by CapeNature, South African National Parks, Eastern Cape Parks and Tourism Agency, National Zoological Gardens, Department of Environmental Affairs, Northern Cape Department of Environment and Nature Conservation, Eastern Cape Department of Economic Development, Environmental Affairs and Tourism and Free State Department of Economic, Small business, Tourism and Environmental Affairs

BMU. (undated) *Bikin Project and Korean Pine Carbon Storage Project: Results of the Russian-German Cooperation in the Russian Far East*. Federal Ministry for the Environment, Nature Conservation, Building and Nuclear Safety of Germany, Germany

Boedhihartono, A.K. (2017) 'Can community forests be compatible with biodiversity conservation in Indonesia?' *Land*, vol. 6, no. 21. Doi:10.3390/land6010021

Bouché, P., Crosmary, W., Kafando, P., Doamba, B., Kidjo, F.C., et al. (2016) 'Embargo on lion hunting trophies from West Africa: An effective measure or a threat to lion conservation?' *PLoS One*, vol. 11, no. 5, p e0155763

Bruner, A.G., Gullison, R.E., Rice, R.E. and da Fonseca, G.A.B. (2001) 'Effectiveness of parks in protecting tropical biodiversity'. *Science*, vol. 291, pp 125–129

Campbell, A., Kapos, V., Lysenko, I., Scharlemann, J.P.W., Dickson, B., et al. (2008) *Carbon Emissions from Forest Loss in Protected Areas*. UNEP World Conservation Monitoring Centre, Cambridge

Carey, C., Dudley, N. and Stolton, S. (2000) *Squandering Paradise?* WWF International, Gland, Switzerland

Caro-Borrero, A., Corbera, E., Neitzel, K.C. and Almeida-Leñero, L. (2015) '"We are the city lungs": Payments for ecosystem services in the outskirts of Mexico City'. *Land Use Policy*, vol. 43, pp 138–148

Carter, N.H. and Allendorf, T.D. (2016) 'Gendered perceptions of tigers in Chitwan national park, Nepal'. *Biological Conservation*, vol. 202, pp 69–77

CBD. (2004) *Programme of Work on Protected Areas*. UNEP and the Secretariat of the Convention on Biological Diversity, Nairobi and Montreal

Ceballos, G., Ehrlich, P.R. and Dirzo, R. (2017) 'Biological annihilation via the ongoing sixth mass extinction signalled by vertebrate population losses and declines'. *Proceedings of the National Academy of Sciences*, vol. 114, no. 30, pp E6089–E6096

Chase, M.J., Schlossberg, S., Griffin, C.R., Bouché, P.J.C., Djene, S.W., et al. (2016) 'Continent-wide survey reveals massive decline in African savannah elephants'. *PeerJ*, vol. 4. https://doi.org/10.7717/peerj.2354

Coad, L., Leverington, F., Knights, K., Geldmann, J., Eassom, A., et al. (2015) 'Measuring impact of protected area management interventions: Current and future use of the global database of protected area management effectiveness'. *Philosophical Transactions of the Royal Society B*, vol. 370, Article Id. 20140281

Conservation Assured. (2018a) *Safe Havens for Wild Tigers: A Rapid Assessment of Management Effectiveness Against the Conservation Assured Tiger Standards*. Conservation Assured, Singapore

Conservation Assured. (2018b) *CA | TS Manual Version 2.0*. Conservation Assured, Singapore

Costanza, R., de Groot, R., Braat, L., Kubiszewski, I., Fioramonti, L., et al. (2017) 'Twenty years of ecosystem services: How far have we come and how far do we still need to go?' *Ecosystem Services*, vol. 28, pp 1–16

Craigie, I.D., Baillie, J.E.M., Balmford, A., Carbone, C., Collen, B., et al. (2010) 'Large mammal population declines in Africa's protected areas'. *Biological Conservation*, vol. 143, no. 9, pp 2221–2228

Creel, S., M'Soka, J., Dröge, E., Rosenblatt, E., Becker, M.S., et al. (2016) 'Assessing the sustainability of African lion trophy hunting, with recommendations for policy'. *Ecological Applications*, vol. 26, no. 7, pp 2347–2357

Damania, R., Seidensticker, S., Whitten, T., Sethi, G., MacKinnon, K., et al. (2008) *A Future for Wild Tigers*. World Bank, Washington, DC

Dudley, N., Belokurov, A., Higgins-Zogib, L., Hockings, M., Stolton, S., et al. (2007) *Tracking Progress in Managing Protected Areas Around the World: An Analysis of Two Applications of the Management Effectiveness Tracking Tool*. WWF, Gland, Switzerland

Dudley, N., Higgins-Zogib, L. and Mansourian, S. (2009) The Links between protected areas, faiths, and sacred natural sites. *Conservation Biology*, vol. 23, no. 3, pp 568–577

Dudley, N. and Stolton, S. (2011) *Death of a Rhino: Lessons Learned from the Disappearance of the Last Javan Rhino in Vietnam*. Equilibrium Research for WWF, Bristol

Dudley, N., Stolton, S., Belokurov, A., Krueger, L., Lopoukhine, N., et al. (2009) *Natural Solutions: Protected Areas Helping People Cope with Climate Change*. IUCN-WCPA, TNC, UNDP, WCS, the World Bank, and WWF, Gland, Switzerland, Washington, DC and New York

Dunn, H. (2010) *Payments for Ecosystem Services*. DEFRA Evidence and Analysis Series, Paper 4. Department of Environment, Food and Rural Affairs, London

Environment Protection Agency. (1999) *Protecting Sources of Drinking Water Selected Case Studies in Watershed Management*. EPA 816-R-98-019, United States Environmental Protection Agency, Office of Water, Washington, DC

Ferraro, P.J. (2009) 'Counterfactual thinking and impact evaluation in environmental policy'. In: Birnbaum, M. and Mickwitz, P. (eds.) *Environmental Program and Policy Evaluation: New Directions for Evaluation*, vol. 22, pp 75–84

Ferraro, P.J. and Pattanayak, S.K. (2006) 'Money for nothing? A call for empirical evaluation of biodiversity conservation investments'. *PLoS Biology*, vol. 4, p e105

Finer, M. and Jenkins, C.N. (2012) 'Proliferation of hydroelectric dams in the Andean Amazon and implications for Andes-Amazon connectivity'. *PLoS One*, vol. 7, no. 4, p e35126

Fisher, M. (2016) 'Fall, resurrection and uncertainty: An Arabian tale'. *Oryx*, vol. 50, no. 1, pp 1–2

Foose, T.J. and van Strien, N. (eds.) (1997) *Asian Rhinos Status and Conservation Plan*. IUCN, Gland, Switzerland

Foxcroft, L.C., Pyšek, P., Richardson, D.M. and Genovesi, P. (eds.) (2013) *Plant Invasions in Protected Areas: Patterns, Problems and Challenges*. Springer, Dordrecht

Franks, P., Small, R. and Booker, F. (2018) *Social Assessment for Protected and Conserved Areas (SAPA). Methodology Manual for SAPA Facilitators*, 2nd edition. International Institute for Environment and Development, London

Funston, P.J., Groom, R.J. and Lindsey, P.A. (2013) 'Insights into the management of large carnivores for profitable wildlife-based land uses in African Savannas'. *PLoS One*, vol. 8, no. 3, p e59044. Doi:10.1371/journal.pone.0059044

Gaskill, M. (2019) 'Scientists: Trophy hunting "not irreplaceable" for conservation funding'. *The Revelator*, 3 December. https://therevelator.org/trophy-hunting-conservation/. Accessed 8 December 2019

Geldmann, J., Coad, L., Barnes, M., Craigie, I.D., Hockings, M., et al. (2015) 'Changes in protected area management effectiveness over time: A global analysis'. *Biological Conservation*, vol. 191, pp 692–699

Geldmann, J., Coad, L., Barnes, M.D., Craigie, I.D., Woodley, S., et al. (2018) 'A global analysis of management capacity and ecological outcomes in terrestrial protected areas'. *Conservation Letters*, vol. 11, p e12434

Golden Kroner, R.E., Qin, S., Cook, C.N., Krithivasan, R., Pack, S.M., et al. (2019) 'The uncertain future of protected lands and waters'. *Science*, vol. 364, pp 881–886

Grima, N. and Singh, S. (2019) 'How the end of armed conflicts influence forest cover and subsequently ecosystem services provision? An analysis of four case studies in biodiversity hotspots'. *Land Use Policy*, vol. 81, pp 267–275

Gross, J.E., Woodley, S., Welling, L.A. and Watson, J.E.M. (eds.) (2016) *Adapting to Climate Change: Guidance for Protected Area Managers and Planners*. Best Practice Protected Area Guidelines Series No. 24. IUCN, Gland, Switzerland

Hanson, T., Brooks, T.M., da Fonseca, G.A.B., Hoffmann, M., Lamoreux, J.F., et al. (2008) 'War in biodiversity hotspots'. *Conservation Biology*, vol. 23, no. 3, pp 578–587

Harihar, A., Chanchani, P., Borah, J., Crouthers, J.R., Darman, Y., et al. (2018) 'Recovery planning towards doubling wild tiger *Panthera tigris* numbers: Detailing 18 recovery sites from across the range'. *PLoS One*, vol. 13, no. 11. Doi:10.1371/journal.pone.0207114

Hockings, M., Adams, W., Brooks, T.M., Dudley, N., Jonas, H., et al. (2013) 'A draft code of practice for research and monitoring in protected areas'. *PARKS*, vol. 19, no. 2, pp 85–94

Hockings, M., Stolton, S., Dudley, N., James, R., Mathur, V., et al. (2008) *Enhancing our Heritage Toolkit Assessing Management Effectiveness of Natural World Heritage Sites*. UNESCO, Paris

Hockings, M., Stolton, S., Leverington, F., Dudley, N. and Courrau, J. (2006) *Evaluating Effectiveness: A Framework for Assessing Management Effectiveness of Protected Areas*, 2nd edition.

Best Practice Protected Area Guidelines no. 14, IUCN and James Cooke University, Gland, Switzerland and Brisbane Australia

Hole, D.G., Willis, S.G., Pain, D.G., Fishpool, L.D., Butchart, S.H.M., et al. (2009) 'Projected impacts of climate change on a continent-wide protected area network'. *Ecology Letters*, vol. 12, pp 420–431

IUCN SULi, IIED, CEED, Austrian Ministry of Environment and TRAFFIC. (2015) *Symposium Report, 'Beyond Enforcement: Communities, Governance, Incentives and Sustainable Use in Combating Wildlife Crime'*. Glenburn Lodge, Muldersdrift, South Africa, 26–28 February

IUCN-TILCEPA. (2010) *Joint PAEL-TILCEPA Workshop on Protected Areas Management Evaluation and Social Assessment of Protected Areas*. IUCN, Gland, Switzerland

Jhala, Y.V., Qureshi, Q. and Nayak, A.K. (eds.) (2019) *Status of Tigers, Co-Predators and Prey in India 2018: Summary Report*. National Tiger Conservation Authority, Government of India, New Delhi and Wildlife Institute of India, Dehradun

Joppa, L.N., Loarie, S.R. and Pimm, S.L. (2008) 'On the protection of "protected areas"'. *Proceedings of the National Academy of Sciences*, vol. 105, pp 6673–6678

Kajala, L. (2012) 'Estimating economic benefits of protected areas in Finland'. In: Kettunen, M., Vihervaara, P., Kinnunen, S., D'Amato, D., Badura, T., et al. (eds.) *Socio-Economic Importance of Ecosystem Services in the Nordic Countries: Synthesis in the Context of the Economics of Ecosystems and Biodiversity (TEEB)*, TemaNord, vol. 559, pp 255–259

Karanth, K.K. and De Fries, R. (2011) 'Nature-based tourism in Indian protected areas: New challenges for park management'. *Conservation Letters*, vol. 4, pp 137–149

Karanth, K.K., Jain, S and Mariyam, D. (2017) 'Emerging trends in wildlife and tiger tourism in India'. In: Chen, J.S. and Prebensen, N.K. (eds.) *Nature Tourism*. Routledge, Oxon

Kettunen, M. (2011) 'Recognising the value of protected areas'. In: ten Brink, P. (ed.) *The Economics of Ecosystems and Biodiversity in National and International Policymaking*. Earthscan, London

Leverington, F., Hockings, M. and Lemos Costa, K. (2008) *Management Effectiveness Evaluation in Protected Areas: A Global Study*. University of Queensland, Brisbane

Leverington, F., Lemos Costa, K., Pavese, H., Lisle, A. and Hockings, M. (2010) 'A global analysis of protected area management effectiveness'. *Environmental Management*, vol. 46, pp 485–498

Lham, D., Wangchuk, S., Stolton, S. and Dudley, N. (2019) 'Assessing the effectiveness of a protected area network: A case study of Bhutan'. *Oryx*, vol. 53, no. 1, pp 63–70

Lindsey, P.A., Balme, G.A., Booth, V.R. and Midlane, N. (2012) 'The significance of African lions for the financial viability of trophy hunting and the maintenance of wild land'. *PLoS One*, vol. 7, no. 1. Doi:10.1371/journal.pone.0029332

Lindsey, P.A., Balme, G.A., Funston, P.J., Henschel, P.H. and Hunter, L.T.B. (2016) 'Life after Cecil: Channelling global outrage into funding for conservation in Africa'. *Conservation Letters*, vol. 9, no. 4, pp 296–301

Lindsey, P.A., Miller, J.R.B., Petracca, L.S., Coad, L., Dickman, A.J., et al. (2018) 'More than $1 billion needed annually to secure Africa's protected areas with lions'. *Proceedings of the National Academy of Sciences*, vol. 115, p 45

Maiden, J. (1905) 'Miscellaneous notes (chiefly taxonomic) on *Eucalyptus*'. *Proceedings of the Linnaean Society of New South Wales*, vol. 29, pp 777–779

Mansourian, S. and Dudley, N. (2008) *Public Funds to Protected Areas*. WWF, Gland, Switzerland, 48 pages

Mascia, M.B. and Pailler, S. (2011) 'Protected areas downgrading, downsizing and degazettement (PADDD) and its conservation implications'. *Conservation Letters*, vol. 4, pp 9–20

Mathur, V.B., Gopal, R., Yadav, S.P. and Negi, H.S. (2014) *Management Effectiveness Evaluation of Tiger Reserves*. Technical Manual No. WII-NTCA/01/2010, p 21. Revised and updated version. WII-NTCA/01/2014, p 25. National Tiger Conservation Authority and Wild-life Institute of India, New Delhi

McCarthy, D.P., Donald, P.F., Scharlemann, J.P.W., Buchanan, G.M., Balmford, A., et al. (2012) 'Financial costs of meeting global biodiversity conservation targets: Current spending and unmet needs'. *Science*, vol. 338, pp 946–949

Mehta, A. and Kill, J. (2007) *Seeing Red? "Avoided deforestation" and the rights of indigenous peoples and local communities*, Fern, Brussels and Moreton-in-the-Marsh

Moodley, Y., Russo, I-R.M., Dalton, D.L., Kotzé, A., Muya, S., et al. (2017) 'Extinctions, genetic erosion and conservation options for the black rhinoceros (*Diceros bicornis*)'. *Scientific Reports*, vol. 7, no. 41417. Doi:10.1038/srep41417

Mrdak, D. (2009) *Environmental Risk Assessment of the Morača River Canyon and Skadar Lake*. WWF MedPo and Green Home, Podgorica

Murray, C.K. (2017) *The Lion's Share? On the Economic Benefits of Trophy Hunting*. A report for the Humane Society International, prepared by Economists at Large, Melbourne, Australia

Nagendra, H. (2008) 'Do parks work? Impacts of protected areas on land cover clearing'. *Ambio*, vol. 37, pp 330–337

Nardelli, F. (2016) 'Current status and conservation prospects for the Javan rhinoceros *Rhinoceros sondaicus* Desmarest 1822'. *International Zoo News*, vol. 63, pp 180–202

Nelson, A. and Chomitz, K. (2009) *Protected Area Effectiveness in Reducing Tropical Deforestation*. The World Bank, Washington, DC

Neugarten, R.A., Langhammer, P.F., Osipova, E., Bagstad, K.J., Bhagabati, N., et al. (2018) *Tools for Measuring, Modelling, and Valuing Ecosystem Services: Guidance for Key Biodiversity Areas, Natural World Heritage Sites, and Protected Areas*. IUCN, Gland, Switzerland

Newmark, W. (1995) 'Extinction of mammal populations in Western North American national parks'. *Conservation Biology*, vol. 9, no. 3, pp 512–526

Nguyen, M.H., Vu, V.D., Nguyen, V.S., Hoang, V.T., Nguyen, H.D., Pham, N.T., Than, T.H. and Doan, C. (2007) *Report on the Review of Vietnam's Wildlife Trade Policy*. CRES/FPD/UNEP/CITES/IUED, Hanoi, Vietnam

Pack, S.M., Napolitano Ferreira, M., Krithivasan, R., Murrow, J., Bernard, E., et al. (2016) 'Protected Area Downgrading, Downsizing and Degazattmenet (PADDD) in the Amazon'. *Biological Conservation*, vol. 197, pp 32–39

Pagiola, S., Bishop, J. and Landell-Mills, N. (eds.) (2002) *Selling Forest Environmental Services: Market-Based Mechanisms for Conservation and Development*. Earthscan, London

Pathak, N. (ed.) (2009) *Community Conserved Areas in India: A Directory*. Kalpavriksh, Pune, India

Pekor, A., Miller, J.R.B., Flyman, M.V., Kasiki, S., Kesch, M.K., et al. (2019) 'Fencing Africa's protected areas: Costs, benefits and management issues'. *Biological Conservation*, vol. 229, pp 67–75

Pimm, S.L., Jenkins, C.N. and Li, B.V. (2018) 'How to protect half of earth to ensure it protects sufficient biodiversity'. *Science Advances*, vol. 4, no. 8. Doi:10.1126/sciadv.aat2616

Polasky, S., Tallis, H. and Reyes, B. (2015) 'Setting the bar: Standards for ecosystem services'. *Proceedings of the National Academy of Sciences*, vol. 112, pp 7356–7361

Porras, I., Barton, D.N., Miranda, M. and Chacón-Cascante, A. (2013) *Learning from 20 Years of Payments for Ecosystem Services in Costa Rica*. International Institute for Environment and Development, London

Poulsen, J.R., Koerner, S.E., Moore, S., Medjibe, V.P., Blake, S., et al. (2017) 'Poaching Empties Critical Central African Wilderness of Forest Elephants'. *Current Biology Magazine*, vol. 27, pp R123–R138

Pyhälä, A., Eklund, J., McBride, M.F., Rakotoarijaona, M.A. and Cabeza, M. (2019) 'Managers' perceptions of protected area outcomes in Madagascar highlight the need for species monitoring and knowledge transfer'. *Conservation Science and Practice*, vol. 1, p e6. Doi:10.1002/csp2.6

Rao, M., Nagendra, H., Shaabuddin, G. and Carrasco, L.R. (2016) 'Integrating community-managed areas into protected area systems: The promise of synergies and the realities of trade-offs'. In: Joppa, L., Baille, J.E.M. and Robinson, J.G. (eds.) *Protected Areas: Are They Safeguarding Biodiversity?* Wiley Blackwell and Zoological Society, London

Reddy, C.S. and Yosef, R. (2016) 'Living on the edge: Attitudes of rural communities toward Bengal tigers (*Panthera tigris*) in Central India'. *Anthrozoös*, vol. 29, no. 2, pp 311–322. Doi :10.1080/08927936.2016.1152763

Redford, K.H. and Fearn, E. (2006) *Ecological Future of Bison in North America: A Report from a Multi-Stakeholder, Transboundary Meeting.* Wildlife Conservation Society, New York

Rich, T.C.G. and Houston, L. (2004) 'The distribution and population sizes of the rare English endemic *Sorbus wilmottiana* E.F. Warburg, Wilmott's Whitebeam'. *Watsonia*, vol. 25, pp 185–191

Riggio, J.S., Jacobson, A., Dollar, L., Bauer, H., Becker, M., et al. (2013) 'The size of savannah Africa: A lion's (*Panthera leo*) view'. *Biodiversity and Conservation*, vol. 22, pp 17–35

Saunders, J. and Nussbaum, R. (2008) *Forest Governance and Reduced Emissions from Deforestation and Degradation (REDD).* Briefing Paper EEDP LOG BP 08/01, Chatham House, London

Save the Rhino (2010) 'Illegal trade in rhino horn: the Vietnamese connection'. 16th August 2010. http://www.rhinoconservation.org/2010/08/16/illegal-trade-in-rhino-horn-the-vietnamese-connection/, accessed January 9th 2020

Schaller, G.B., Nguyen, X.D., Le, D.T. and Vo, T.S. (1990) 'Javan rhinoceros in Vietnam'. *Oryx*, vol. 24, pp 77–80

Scherr, S.J., Bennett, M.T., Loughney, M. and Canby, K. (2006) *Developing Future Ecosystem Service Payments in China: Lessons Learned from International Experience.* Forest Trends, Washington, DC

Selulić, G., Ivanić, K.Z. and Porej, D. (2017) *Protected Areas Benefits Assessment (PA-BAT) in Montenegro.* WWF Adria, Zagreb

Soil Association. (1990) *Guidelines for Conservation.* Soil Association, Bristol

Spergel, B. (2001) *Raising Revenues from Protected Areas: A Menu of Options.* WWF-US, Washington, DC

Stolton, S. and Dudley, N. (2009) *The Protected Areas Benefits Assessment Tool.* WWF, Gland, Switzerland

Stolton, S. and Dudley, N. (2016) *METT Handbook: A Guide to Using the Management Effectiveness Tracking Tool (METT).* WWF-UK, Woking

Stolton, S., Dudley, N., Belokurov, A., Deguignet, M., Burgess, N.D., et al. (2019) 'Lessons learned from 18 years of implementing the management effectiveness tracking tool (METT): A perspective from the METT developers and implementers'. *PARKS*, vol. 5, no. 2, pp 79–92

Stolton, S., Hockings, M., Dudley, N., MacKinnon, K. and Whitten, T. (2003) *Reporting Progress in Protected Areas: A Site Level Tracking Tool.* WWF and the World Bank, Gland and Washington, DC

Stoner, S.S. and Pervushina, N. (2013) *Reduced to Skin and Bones Revisited: An Updated Analysis of Tiger Seizures from 12 Tiger Range Countries (2000–2012).* TRAFFIC, Kuala Lumpur, Malaysia

Sulyandziga, P. (2017) 'Parks and arbitration: A leader of Russia's Udege community describes the decades-long fight to create Bikin national park, the first to safeguard indigenous rights'. *World Policy Journal*, vol. 34, pp 6–10

UICN/PACO. (2009) *La grande chasse en Afrique de l'Ouest: quelle contribution à la conservation?* (*Big Game Hunting in West Africa: What Is Its Contribution to Conservation?*) IUCN PACO, Ouagadougou, Burkina Faso

UNDP and GEF. (2011) 'The economic value of protected areas in Montenegro'. Quoted in: Kettunen, M. and ten Brink, P. (eds.) *Social and Economic Benefits of Protected Areas: An Assessment Guide.* Earthscan, London

UNEP-WCMC and IUCN. (2018) *2018 United Nations List of Protected Areas: Supplement on Protected Area Management Effectiveness.* UNEP World Conservation Monitoring Centre, Cambridge

UNESCO. (2017) Nomination Document for the Bikin River Valley as an Extension to the Central Sikhote-Alin World Heritage Property

Veldhuis, M.P., Ritchie, M.E., Ogutu, J.O., Morrison, T.A., Beale, C.M., et al. (2019) 'The Serengeti squeeze: Cross-boundary human impacts compromise an iconic protected ecosystem'. *Science*, vol. 363, no. 6434, pp 1424–1428. Doi:10.1126/science.aav0564

Verma, M., Negandhi, D., Khanna, C., Edgaonkar, A., David, A., et al. (2015) *Economic Valuation of Tiger Reserves in India: A Value+ Approach.* Indian Institute of Forest Management. Bhopal, India

Verma, M., Negandhi, D., Khanna, C., Edgaonkar, A., David, A., et al. (2017) 'Making the hidden visible: Economic valuation of tiger reserves in India'. *Ecosystem Services*, vol. 26, pp 236–244

Vo Quy. (2005) *The Attack of Agent Orange on the Environment in Vietnam and Its Consequences.* Agent Orange and Dioxin in Vietnam 35 Years Later: Proceedings of the Paris Conference Senate, 11–12 March. France-Vietnam Friendship Association

Vujović, A., Krivokapić, Z., Stefanović, M., Pešić, V. and Jovanović, J. (2018) 'Integrated lake basin management for lake Skadar/Shkodra'. In: *The Handbook of Environmental Chemistry.* https://link.springer.com/chapter/10.1007/698_2018_264

Walston, J., Robinson, J.G., Bennett, E.L., Breitenmoser, U., da Fonseca, G.A.B., et al. (2010) 'Bringing the tiger back from the brink – the six percent solution'. *PLoS Biology*, vol. 8. Doi:10.1371/journal.pbio.1000485

Wolf, C. and Ripple, W.J. (2017) 'Range contractions of the world's large carnivores'. *Royal Society Open Science*, vol. 4. http://dx.doi.org/10.1098/rsos.170052

World Rainforest Movement and Grain. (2014) *REDD Alert! How REDD+ Projects Undermine Peasant Farming and Real Solutions to Climate Change.* World Rainforest Movement and Grain, Montevideo and Barcelona

WWF. (2017) *Beyond the Stripes: Save Tigers, Save so Much More.* WWF International, Gland, Switzerland

Wyborn, C., van Kerkhoff, L., Dunlop, M., Dudley, N. and Guevara, O. (2016) 'Future oriented conservation: Knowledge governance, uncertainty and learning'. *Biodiversity Conservation.* Doi:10.1007/s10531-016-1130-x

11
THE COSTS

Protection doesn't come free. Along with the benefits, there are associated financial costs in terms of the money needed to establish and manage protected areas and other forms of area-based conservation. Costs also include any benefits forgone by those who would have used the available land and water for something else. More controversially, there are also costs in terms of human societies. There is persistent criticism of protected areas from some human rights groups, particularly relating to dispossessing people from their traditional territory and resources, treatment of indigenous peoples and local communities and human-wildlife conflict. Many protected areas have been established on long-established lands and waters of indigenous people or other local communities. These people may in consequence have lost access to natural resources and to culturally or spiritually significant sites or even been expelled from the area altogether, creating situations that have become known as "green grabbing" and conservation refugees. The scale and seriousness of these side effects are examined, along with some of the responses and whether or not they have gone far enough in addressing human rights concerns.

Global meetings like IUCN's World Conservation Congress are places where ideas are shared, debates take place and new contacts are made between people from different countries, disciplines and opinions. Or are they? Our own experience is that what we might call the "human rights" and the "nature conservation" groups stick to their own rooms, their own work streams, and that their conversations have little in common. For the few people who flit between the two, it can seem like entering different worlds. And while both worlds tend to focus a lot on threats, in one the threat tends to be still mainly perceived as from "people against nature" and in the other it is increasingly the "threats that conservation poses to people". There are two parallel streams evolving in the conservation arena, splitting effort, promoting infighting and sometimes losing sight of the larger objectives in consequence.

In one world, protected areas are viewed as the single most important conservation tool available at the moment. In the other world, protected areas are viewed as an almost wholly negative imposition on people, which should be abandoned as a historical mistake. Both claim the moral high ground.

Following the 1972 Stockholm Environment Conference, popularly regarded as marking the birth of the modern environmental movement, conservation was seen as a generally "good" thing in the mass of public opinion, at least in the richer countries. This remains true today. As discussed, with recognition of the scale of threats facing wildlife, conservation bodies and some governments raced to set up national parks and nature reserves, mainly based on models already applied in the United States, Australia and India. Continuing pressures such as poaching and incursions into protected areas were generally dismissed as criminal acts to be prevented; their instigators faceless miscreants without a back story. For several decades, at least within the broad sweep of what became known as the environmental movement, this situation remained the same. Then in the late 1990s and early years of the twentieth century we started to hear a different perspective (e.g. Gray et al., 1998; Nelson and Hossack, 2003; Colchester, 2003), of people being dispossessed of their ancestral lands or being instantly criminalised for following ways of life that had been in place for centuries but which some faraway government had now ruled to be illegal. Stories started to emerge of people being punished for taking food from reserves that they desperately needed to feed their families, in much the same way as people were hanged for poaching in the aristocracy's hunting grounds in Medieval Europe. At the Fifth World Parks Congress in 2003, in the edgy city of Durban, South Africa, a well-organised caucus of indigenous people demanded that conservationists take human rights more seriously, extracting some important commitments relating to conservation and human rights. In the same way that the Stockholm Conference shook up the world in 1972, for many of us the Fifth World Parks Congress marked another fundamental shift in approach.

Conflicts in Cameroon

In 2018 and 2019, the indigenous peoples' rights group Survival released reports accusing conservation organisations of funding local forest guards who were guilty of persistent racism including physical attacks and rape against baka (pygmy) and other local communities in both Cameroon (Pyhälä et al., 2016) and the Democratic Republic of Congo. The resulting publicity saw accusations of cover-ups, members of the German government questioning whether to withdraw conservation funding, and a huge soul-searching by conservation NGOs. I have been to some of the places mentioned – Dja and Lobeké National Parks – and met with the baka, although not for a long time. Some of the things mentioned in the report did not match my experience. I stayed in the tented camp that served as a project headquarters in Lobeke, where the baka people were working happily alongside other local rangers; in the evenings we ate together, and everything seemed relaxed. But I talked to someone recently who had visited and said the camp had fallen into disrepair. I also talked with Stephen Correy, the director of Survival (I'm a long-standing member) and to a number of other people who know the area better and more recently than I do. I followed and contributed to an email stream from the ICCA Consortium and was also passed another email stream from a more conservation-focused discussion. And amongst a large group of intelligent and committed people there was a huge divergence of views. When we talked in detail about a particular place and problem, issues did not fall neatly into "conservation" and "human rights" camps. Everyone agreed there were problems, influenced by ancient tensions between the various people living in the area, and there was a range of opinions about what might realistically be done about it. How much can an outside organisation do to address long-standing inter-communal hostilities? Are there circumstances where walking away is the only option? Could conservation be a positive spur to improvement? **ND**

We believe that a global protected area network is a fundamental building block of a sustainable planet. But some of the staunchest critics of protected areas are people whose views we understand; they could have been ours in different times, different lives. And many, probably most, critics of protected areas want much the same end-result as the more traditional conservation lobby, but they disagree about strategy and tactics. How did we get into this mess? And how do we get out of it? These are some of the things that we want to explore in the following chapter.

On the one hand, new thinking about conservation has led to some substantially new approaches, where conservation bodies, governments, indigenous people and local communities have worked together to find mutually acceptable ways forward to balance conservation and development (e.g. Phillips, 2003). These new approaches to conservation now cover many millions of hectares of the planet's surface (see some of the examples in Stevens, 2014). But this hasn't solved all the problems, and criticism continues to emerge (e.g., Chapin, 2004) – a steady trickle of papers in academic journals, reports from activists, newspapers, blogs and exposés. Some human rights bodies and independent researchers have made conservation a particular focus of their activities. Criticism has also come further into the mainstream. In recent years the UN Commissioner on Human Rights has published reports on protected areas and human rights for instance, both sharply critical of the conservation lobby (e.g. Tauli-Corpuz, 2016).

What the critics say

Critics of protected areas are concerned with the number of cases where people – often indigenous people, poor people or other socially disadvantaged and relatively powerless local communities – have been forcibly removed from their traditional lands to create protected areas, or have lost access to natural resources, access to important sacred sites, migration routes and traditional grazing. There are associated concerns about the treatment of local communities by protected area staff. They suggest that this is a larger and more serious human rights issue than has generally been perceived, with millions of people suffering life-changing impacts as a result of conservation in protected areas (e.g. Dowie, 2009). Furthermore, they say that Western concepts of "wilderness" and untouched nature are largely fantasy, which ignore the fact that most supposedly wild places have been inhabited for millennia, and that removing the influence of traditional management can do more harm than good to ecosystems (Cronon, 1996). And finally, critics of protected areas point out that despite a dramatic increase in the amount of land and water under protection, biodiversity is still rapid declining (e.g., Pimm et al., 2014), suggesting that this conservation model doesn't work anyway.

There are therefore three key issues of concern: do protected areas actually protect biodiversity? Is their success helped or hindered by removing people? And do protected areas damage human rights? None of these have a neat black-and-white answer.

Do protected areas protect biodiversity?

We have examined effectiveness in detail in Chapter 10. But if protected areas are ineffective conservation tools, the rest of the conversation is somewhat redundant, so a quick summary follows. Strangely, given the commitment of land and resources to protected areas, until recently we didn't have much more than hearsay and personal experience to go on in judging overall effectiveness. But following the work of an IUCN World Commission on Protected Areas task force, data gathering by the Zoological Society of London and a mass of individual or collective research projects, we can now state confidently that well-managed protected areas do conserve ecosystems and species far more effectively than most other management regimes. Many rare species are now confined entirely within protected areas or are heavily dependent on protected areas for their survival. Other species, now no longer in danger of extinction, owe their survival to protected areas in the past. The current protected area estate has not been enough to stop the loss of biodiversity in its tracks, but it has almost certainly slowed it down very substantially, and we would be in a much worse state without protected areas.

Is the success of protected areas helped or hindered by removing people?

Traditional protected area management assumed that a national park or nature reserve was necessarily an area of land or water devoid of humans. This is still enshrined in a number of countries' protected area legislation, causing all kinds of problems. There has been a backlash against this model in recent years, pointing out that humans have lived alongside other species in most of the world's richest biodiversity areas, co-evolving over hundreds or thousands of years. Removing people might in many cases also remove the types of low-level management needed to maintain this biodiversity. There are certainly situations where humans and biodiversity survive in harmony together and there is long-standing work on conservation in these situations, for instance in the Mediterranean. But there are other cases where they do not, often because conditions change; for example, there are more humans, or people living in the area change the way they manage natural resources, or the environment changes in ways that undermine sustainable management. Many protected areas were set up precisely because existing management was destroying ecosystems or threatening species; utopian claims of local people living in harmony with nature don't always stack up. There is no single or simplistic answer to the question at the top of this paragraph. Some ecosystems indeed can cater for or even require human presence; others may not, or may benefit from decreased human activity.

What we can say with confidence is that removing people from an area seldom works successfully as conservation, let alone considering the human rights implications, if it leaves a legacy of angry, disaffected people behind.

Do protected areas damage human rights?

Again, there is no black-and-white answer, despite claims and counter claims. There have definitely been cases where people have been dispossessed; we have visited some of these and spoken to the people concerned. This generally happens in places in countries where human rights are in general poorly protected. There have also been cases where there were clear agreements at the time, but "dispossession" has emerged as an issue long afterwards for completely unconnected reasons. Some displacements happened so long ago, or were part of much wider changes in ownership, that it is hard to see what can be done about it after this time; the whole colonial period throughout the tropics resulted in vast areas of land being stolen from the original inhabitants. But there are also cases where the issue is still very definitely "live" and needs to be addressed. This is the most important of the three questions we posed, and we therefore look at it in greater detail in the following.

Dispossessing people of their territories and resources

Outside Europe, many of the earliest "modern" protected areas were set up under a colonial or quasi-colonial system, where the indigenous inhabitants had little say in decisions about natural resource management. Early national parks in Australia and the United States were established on land originally belonging to indigenous people, as were countless farms, mines, towns etc. This situation continued throughout the early days of modern protected area expansion and in some countries, it continues today; in India for example the law says people cannot live in national parks and a new national park generally means relocating people.

Nyika National Park, Malawi

The exclusionary approach to conservation continued after independence in many former colonial countries. Nyika National Park, IUCN Category

II, covers 3,134 km² in northern Malawi, bounded by the Rift Valley, Lake Malawi and the Luanga Valley. Some 940 km² of the area is high plateau above 1,800 metres. The lower slopes are *Brachystegia* woodland and most of the upland plateau is short grass, with small patches of evergreen woodland and fragments of juniper forest (*Juniperus procera*), the southernmost limit of this species. The catchment supplies 40 per cent of water for northern Malawi. Fragments of juniper forest have been protected since the 1930s, and the current borders were set in 1978. The park has high levels of endemism, across a wide range of groups including small mammals.

Before the park was established around 5,000 people lived in the area, practising a mixture of hunting and farming. A relatively low level of settlement has been in place for thousands of years. When Malawi's autocratic former president, Hastings Banda, set up the park in 1978, people were expelled. Not only did they lose their land, but they were forced off the high plateau, which was too cold for malarial mosquitoes, into the lowlands. Without acquired resistance, many are said to have died as a result of catching malaria. The land they were relocated to was also said to be less productive land, causing further resentment. The protected area managers were left with many problems to try to resolve. Attempts to address these issues included establishment of natural resource committees, beekeeping clubs and a revenue sharing scheme. Local people are allowed to place beehives in the park, generating around 8 tonnes of honey a year, and to collect medicinal plants, wild fruit and grass, which are monitored. On application, communities can also use a range of sacred sites (a waterfall, mountain and lake) in rain-making ceremonies. Poaching has also been a major problem, of game, orchid tubers and timber, leading to a dramatic decline in some species. Illegal fire setting was used to maintain open habitat for game or sometimes in revenge, for instance if a poacher was arrested. When I visited a decade ago, 400–500 antelope were being poached a year. I was invited to lunch with the three traditional chiefs of the land covered by the protected area, something that would apparently not have happened a few years earlier. They had generally come to terms with the existence of the protected area and were supportive of plans (that never materialised) to have the site declared as a World Heritage area. But there is no way that the establishment of the protected area was done in a fair or equitable manner. **ND**

To what extent is this standard? It's hard to say, although people have tried. Some estimates suggest for instance that 14 million "conservation refugees" have been created in Africa since the beginning of modern conservation in the nineteenth century and a million people have been displaced in the creation of tiger reserves in India (e.g., Torri, 2011). Writing in 2009, Mark Dowie (2009) said that of the then roughly 100,000 protected areas established since 1900, roughly half were on the territories of indigenous people, many of whom had been expelled. In our experience, a more usual situation is that they have "officially" been expelled but continue to live in the area and estimates in the 1990s for instance were that 80 per cent of Latin American protected areas contained substantial human populations (Amend and Amend, 1995). Sometimes this is with the tacit agreement of protected area mangers, sometimes not. It should also be noted that we have also come across circumstances where what was actually happening on the ground was not reported nationally. For many protected area managers, the legal system under which they have to work is far from supportive: not strong enough to deter illegal activities but far too prescriptive in terms of local communities living or accessing the protected area. We have met managers who have come up with innovative solutions, such as "invasive species control" which allows local people to continue to access fuelwood as long as they take non-native species in areas where all access is in theory strictly forbidden. But such activities were never mentioned in management plans or in reporting in case of additional sanctions against such activities.

Top-down policies towards protected area establishment generally do not exist in isolation; if a country carries out conservation in a repressive, unequal fashion it will generally be doing the same in other situations as well. Many of the largest protected areas have been established in areas without many people, or where people have been driven out by causes other than wildlife conservation. Some of the largest protected areas in eastern Africa are in places where tse tse fly infestation makes cattle rearing difficult. The Tsavo area in Kenya had no permanent settlements largely because of regular incursions by Somali raiders. Sometimes land claims emerge long after the protected area is in place and may or may not be legitimate. When we talk to people who were involved in setting up protected areas in the 1960s and 1970s, they often recall long negotiation processes that seem to have been forgotten in the years since. In countries with no official land title, civil unrest, shifting boundaries and large-scale human movement, working out who owns what can be incredibly challenging. Outside of protected area management, we worked with a company in Borneo that had spent months mapping local land ownership patterns over an area of land slated for plantation establishment; when they restarted negotiations a few months later many of the boundaries had changed, through death, migration or because rights to land and resources shifted regularly between families during the year. The setting-up of protected areas assumes borders can be mapped out and physically demarcated. But in many countries these borders are still disputed or unknown. This should not stop area-based conservation, but it certainly makes establishment and management much harder when land and water ownership and boundaries are obscure.

Choosing sparsely populated areas makes things much easier from a social point of view; most managers of large wilderness areas simply come to some arrangement with remaining human inhabitants, whatever the government says. But this has also pushed conservation efforts towards the margins where there is also sometimes little biodiversity, resulting in accusations that protected areas are missing out on the most important biodiversity (see Chapter 7). Damned if you do, damned if you don't.

Every square metre of farmland, commercial forestry, reservoir and urban area in many countries has been taken, usually by force, from the original inhabitants.

People rarely debate about whether ranchers should give back their holdings to the people their ancestors forced off the land a few hundred years ago in countries like the United States or Canada, yet these issues constantly arise with respect to protected areas. And who has the rights? If some of the protected areas in East Africa were to be handed back to the original owners, would these be the Masaii who now claim the land or the Han who were forced off by the Masaii not long before? These are morally complex and extremely sensitive issues that we mention with trepidation; some readers will react with outrage to the last few sentences.

But we need to start having a far more open dialogue than has been the case until now. The tragedy of the current situation is that most indigenous people and traditional communities want much the same end result as most conservationists; or at least their aims are pretty much compatible. Yet we've ended up all too often on opposing sides. This is politically extremely dangerous: two movements that should be pulling together against powerful vested interests instead fighting among ourselves. We need to understand how this has happened, why it has happened and what we can do to fix it. In his detailed study of the situation, Mark Dowie (2009) pleads that

> as conservationists and native people converge uneasily they can come to agreement that they both own the interdependent causes of biodiversity conservation and cultural survival, that they need each other badly, and that together they can create a new conservation paradigm.

We completely agree. There have definitely been some important steps towards this in the last two decades, but there is also undoubtedly still some way to go.

Top-down decisions in Guatemala

Laguna Lachua in Guatemala illustrates some of the complexities. We tested out a community-based approach to forest quality assessment there many years ago. Laguna Lachua is a relatively small but very beautiful rainforest remnant, with a near-circular karstic lake in the middle. We ate black beans and eggs in the park's camp, had to sew up holes in our mosquito net in the evening and, when leaving on the small commercial flight the pilot handed Sue the controls for about fifteen minutes, something that probably wouldn't happen today. If the rumours are true, the protection of the area came about because a woman had a chat to a senior army officer at a cocktail party in Guatemala City. He obligingly drew a block on a map around the lake in what was then unbroken forest and created a national park. There was no consultation. Then in the years since, the whole area has been deforested by settlers, so that today, apart from a small corridor in the south, the national park stands out on Google Earth as a weird square in the middle of an otherwise transformed landscape. And it has also become a classic landscape of contention. For the few remaining local indigenous people, the national park contains an important sacred lake and the only place where conditions still resemble their traditional lands. For conservationists, it contains rare species and remnant rainforest that is becoming increasingly scarce in Central America. And the settlers hate the park, seeing it as wasted land. The day after we were there park rangers were barricaded in their office by angry people demanding access and things have got worse in the twenty years since (Bullock et al., 2019). Illegal logging is rife, and those who protest are threatened.

What does an example like this mean? The process was terrible, but if that late-night party conversation hadn't taken place there would be no rainforest in the area at all, no-where for the indigenous people to aspire to and incidentally the ecosystem services associated with the lake would have been largely squandered. Often in land and water management we have to work with what we have been handed, whether it is national boundaries drawn in ridiculous places by drunk colonial administrators or imperfect land management systems. Trying to straighten things out after the fact is never going to be easy. **ND/SS**

The legal and quasi-legal regime

In theory, over the last fifteen years protected areas have developed a set of social safeguards stronger than almost any other comparable change in land use, enshrined in international agreements and many national laws and policies. The CBD requires Free Prior and Informed Consent before the establishment of protected areas in or near where indigenous people are living, in other words that people give consent for a protected area only once they have been fully apprised of all the facts and if necessary have adequate compensation mechanisms in place.

Most large conservation NGOs and most governments have developed human rights policies related to conservation in protected areas. Many NGOs are currently updating grievance procedures in light of recent bad publicity, and IUCN's Green List of Protected and Conserved Areas stresses governance equality as a fundamental of good management. Criticism of NGOs focuses on: how well they follow their own policies (Newing and Perram, 2019); how much they go along with governments that practise top-down or inequitable approaches to conservation; if they close their eyes to human rights abuses by groups that they fund; who they take funding from and what the funders track records are in human rights; and the extent to which historical grievances should be redressed.

Forced relocations are rejected by the CBD and by IUCN in its own definition of a protected area: "The definition and categories of protected areas should not be used as an excuse for dispossessing people of their land", a clause that was overwhelmingly supported by IUCN members (Dudley, 2008). People can and are forcibly relocated for many reasons around the world: to build roads, hospitals, shopping malls and sports stadiums; to accommodate pipelines; for port facilities, airports and forestry plantations; and, increasingly, to build hydropower schemes. Around half a million people were reported to have been relocated for the 2008

Olympics in Beijing; the Chinese government claims a lower number. People were moved from their family homes, despite their vehement protests, to make way for a new supermarket in the nearest town to us in Wales only five years ago. In theory this should no longer be possible in protected areas. But we all know that governments often don't follow international norms, or their own laws, in the management of natural resources.

If they are done correctly, protected areas can also protect human communities, alleviate poverty and bring benefits to parts of the world where few other options exist (Dudley et al., 2008). Indeed, while individual case studies have certainly demonstrated links between individual protected areas and increased poverty, others have found that protected areas can boost incomes, and several recent meta-analyses have found little evidence of these links on a global scale (e.g., Mammides, 2020).

Conclusions

There are clearly many important social issues that still need to be redressed. One of the first is to start some proper conversations, not only between conservationists and indigenous people and local communities, but also between the two "sides" within the conservation community that we identified at the start of this section. There are already some important areas of agreement: that large areas of the planet should remain as natural or near natural ecosystems, that other species have existence rights and that territories of indigenous people are often important sanctuaries for biodiversity. There is considerable but not universal agreement that protected areas are a necessary part of the conservation estate and that they supply other important ecosystem services. There is however far less agreement on the current efficiency of protected areas as tools for conservation, the extent to which they now meet human rights concerns, the need for reparation for historic human rights violations and whether conservation needs should ever trump human needs, and if so what this means in practice.

Stepping back to our examples in the text, there are extremely good reasons to protect Nyika Plateau if it supplies almost half the water for the north of the country, let alone its wildlife values. But could there have been another way of doing this, which had less of an impact on the small number of people who lived there? Clearly, yes. Is Laguna Lachua worth preserving? Again, we would say yes, both because it is clearly a sacred site for the embattled local community and for its watershed and wildlife values; in fact, the protected area should be far larger. But a different approach would have helped: it is not hard to envisage the original inhabitants standing shoulder to shoulder with the park rangers rather than attacking the site if things had been handled differently.

Before setting up a protected area that restricts local peoples' access and resource use, there therefore needs to be a full and participatory analysis of the conservation problem that the protected area aims to solve: its severity, causes, the positive or negative role played by local communities and what is needed to help address the conservation problem (there may be more than one option). This will then help determine whether any social impacts are proportionate to the scale of the conservation challenge (Newing and Perram, 2019). Social impact assessments need to address issues such as perceptions of these impacts on individuals directly affected by protected areas, understanding both actual and perceived impacts and recognising the dynamic nature of many of these impacts (Jones et al., 2017). We will return to some of these issues in the final section of the book, investigating where we go from here.

References

Amend, S. and Amend, T (eds.) (1995) *National Parks Without People? The South American Experience*. Parques Nacionales y Conservacion Ambiental no. 5, Quito

Bullock, E.L., Nolte, C., Segovia, A.R. and Woodcock, C.E. (2019) 'Ongoing forest disturbance in Guatemala's protected areas'. *Remote Sensing in Ecology and Conservation*. Doi:10.1002/rse2.130

Chapin, M. (2004) 'A challenge to conservationists'. *Worldwatch Magazine*, November–December

Colchester, M. (2003) *Salvaging Nature: Indigenous Peoples, Protected Areas and Biodiversity Conservation*. World Rainforest Movement and Forest Peoples Programme, Montevideo and Moreton-in-Marsh

Cronon, W. (1996) 'The trouble with wilderness, or getting back to the wrong nature'. *Environmental History*, vol. 1, no. 1, pp 7–28

Dowie, M. (2009) *Conservation Refugees*. The MIT Press, Cambridge, MA

Dudley, N., Mansourian, S., Stolton, S. and Suksuwan, S. (2008) *Safety Net: Protected Areas and Poverty Reduction*. WWF International, Gland, Switzerland

Gray, A., Parellada, A. and Newing, H. (eds.) (1998) *Indigenous Peoples and Biodiversity Conservation in Latin America*. IWGIA Document no. 87, International Work Group for Indigenous Affairs, Copenhagen

Jones, N., McGinlay, J. and Dimitrakopoulos, P.G. (2017) 'Improving social impact assessment of protected areas: A review of the literature and directions for future research'. *Environmental Impact Assessment Review*, vol. 64, pp 1–7

Mammides, C. (2020) 'Evidence from eleven countries in four continents suggests that protected areas are not associated with higher poverty rates'. *Biological Conservation*, vol. 241, no. 108353. Doi:10.1016/j.biocon.2019.108353

Nelson, J. and Hossack, L. (eds.) (2003) *Indigenous Peoples and Protected Areas in Africa*. Forest Peoples Programme, Moreton in the Marsh

Newing, H. and Perram, A. (2019) 'What do you know about conservation and human rights?' *Oryx*, vol. 53, no. 4, pp 595–596

Phillips, A. (2003) 'Turning ideas on their head: The new paradigm for protected areas'. *George Wright Forum*, vol. 20, no. 2, pp 8–32

Pimm, S.L., Jenkins, C.N., Abell, R., Brooks, T.M., Gittleman, J.L., et al. (2014) 'The biodiversity of species and their rates of extinction, distribution and protection'. *Science*, vol. 344, no. 6187, p 987

Pyhälä, A., Osuna Orozco, A. and Counsell, S. (2016) *Protected Areas in the Congo Basin: Failing Both People and Biodiversity?* Rainforest Foundation UK, London

Stevens, S. (ed.) (2014) *Indigenous Peoples, National Parks and Protected Areas: A New Paradigm Linking Conservation, Culture and Rights*. University of Arizona Press, Tucson

Tauli-Corpuz, V. (2016) *Report of the Special Rapporteur of the Human Rights Council on the Rights of Indigenous Peoples*. United Nations A/71/229. http://unsr.vtaulicorpuz. org/site/index.php/en/documents/annual-reports/149-report-ga-2016. Accessed 15 November 2019

Torri, M.C. (2011) 'Conservation, relocation and the social consequences of conservation policies in protected areas: Case study of the Sariska Tiger Reserve, India'. *Conservation and Society*, vol. 9, no. 1, pp 54–64

PART IV

Where do we go from here?

12

LEAVING SPACE FOR NATURE

What next?

Finally, some thoughts for the future. Cautious optimism needs to be balanced by a realistic appraisal of the challenges facing conservation. We identify six necessary trends. (1) From protected areas to area-based conservation; a dramatic widening of the tools used in conservation to bring larger areas into the conservation estate. (2) From management to stewardship; a similar broadening of the people involved in area-based conservation. (3) Recognising all the benefits; identifying area-based conservation as a major vehicle, perhaps the major vehicle, for delivering ecosystem services. (4) Connecting through the mosaic; developing area-based conservation systems rather than isolated protected areas. (5) Focusing on delivery; through greater effectiveness and capacity building. (6) Restoring the Earth; we are no longer just talking about "saving" nature but rebuilding functional ecosystems in the many parts of the world where they have become degraded or destroyed. We end the book on a note of cautious optimism.

Area-based conservation is on a cusp at the moment; there is a steady and powerful body of opinion to set aside far larger areas than has been the case until now and a recognition of the critical role of natural ecosystems in maintaining a balanced and healthy planet. The half earth movement is gathering momentum; many people with little interest in conservation feel instinctively that to convert more than half the planet's land surface to agriculture and urban areas is to be avoided. Climate change will create dramatic changes in the way that we live, do business and plan our futures. The bad news about biodiversity loss is getting a higher profile than ever. Setting aside land and water for nature is an intrinsic part of this movement.

But this message for action is far from universally agreed. There has been a dramatic rise in populism, almost all of which is anti-environmental in message. In cases like the recent felling of parts of Białowieża National Park in Poland, the largest ancient forest left in Europe, degradation seems a deliberate act of provocation, a brutal restatement of the old view that nature was made for "man" to exploit with no independent rights. As we write, anger is swelling about increased burning in the Amazon and incursions into indigenous peoples' lands, with positions hardening on both sides. This is one of the toughest times to be engaged in the practice of conservation.

Protected areas, OECMs and other management units do not exist in a vacuum, cut off from the rest of the world; political, social and economic changes all affect them. Crystal ball gazing is difficult, and projections are often wrong. But without a cataclysm of biblical proportions the world will in the future contain more people, with different expectations, fewer resources available and a more unstable climate. We summarised these pressures in Chapter 8. Many old ways of life will wither and, if not actually disappear, become less significant. More people could well mean less land for conservation. Urbanisation and agricultural abandonment could on the other hand mean more land for conservation through land abandonment. It is always hard to read the future, perhaps particularly so at the moment.

Dramatic changes in the far north of the USA

Kenai Fjords National Park, Alaska, covers 2,711 km^2 of the Kenai Peninsula in south-central Alaska, near the town of Seward. The park contains the Harding Icefield, one of the largest in the United States, source of around thirty-eight glaciers that have carved a series of spectacular fjords into the coastline. Most of the area is only accessible by boat, seaplane or by hiking; the only road leads to Exit Glacier not far from Seward. The park is rich in wildlife, with moose, sea otter, humpback whales, orcas and a variety of Arctic birds such as the tufted and horned puffins and murres. We were hosted by the US National Parks Service for a meeting about climate change and protected areas in summer 2017 and flew in by small plane over the vast icefield that has for millennia been a permanent feature in the landscape; the Harding Icefield is around 80 km long and 50 km wide, covering an area of around 18,000 km^2. We also went along the coast by boat and saw the end of one of the glaciers, towering above us in a vast wall of ice. Looking at my photographs the ice wall appears to be about ten times higher than the tour boat sitting in front, the latter itself being the height of a two- or three-storey house. But that must be a massive underestimate; tour boats keep well away from the glacier to avoid falling ice, so my own view is foreshortened. Yet the glacier has been retreating since the 1950s and the rate of loss is accelerating (Hall et al., 2005). It may look immovably large now, but scientists can calculate its likely date of disappearance. Due to complex climatic patterns, Alaska is making the world's largest contribution to sea-level rise from glacier melt, despite only holding 1 per cent of global ice. Shoreline retreat in some regions is averaging 4–8 metres a year due to less ice protection of the shoreline, rising sea levels and permafrost melt. At Exit Glacier the observation site originally built to see the end of the glacier in the 1980s is now surrounded by dense young forest and painted signs mark dramatic year by year retreats. And warming is already making some huge changes to ecology. National

Park staff told us that between 2013–2016 an unusual warming period meant that around a hundred thousand common murres (*Uria aalge* also known as the common guillemot) died as a result of warmer water forcing the fish they feed on down to deeper waters; this loss was compounded by a year of zero breeding success. Losses are continuing (National Park Service, 2019). Coastal lagoons that were freshwater are now becoming brackish as sea levels rise. Stella's eider duck relies on ice for breeding and stands to lose habitat as the ice retreats. Ice retreat also means that the once-mythical Northwest Passage is now a reality and attracting both commercial shipping and tour boats, with a 500 per cent increase in vehicles going through the Bering Strait, creating hazards for the bowhead whale (*Balaena mysticetus*). This is a particularly poignant issue for us; one of our earliest projects was on threats to the bowhead whale (Dudley and Gordon Clarke, 1983) and since then the local Inuit people have done a fantastic job of sustainably managed hunting such that the population has been rebuilt, only now to be threatened again by tourism. These changes do not only affect wildlife values; fish and game are critical issues of food security for many people living in the region. Ocean acidification in particular is a critical threat to many fish species used by native peoples throughout the region (Lynn et al., 2013). Protected area staff face strictly limited options about how they might try to manage under such circumstances; the changes are so vast and dramatic that there may be little to do except monitor what is happening. **ND**

A short reality check: in light of the preceding, is talk of major advances in area-based conservation just a fantasy? The world is undergoing an unprecedented rate of development expansion. To pick just one example, growth in linear infrastructure supports the "roads lead to prosperity" view of development. We have seen the once dust-covered roads through villages leading to Ngorongoro Conservation Area and Serengeti National Park in Tanzania be tarmacked, increasing pressures on conservation but improving the lives of villagers. And we watched in awe as workers, balancing like trapeze artists, fixed cables on transmission towers in Bhutan, cutting swathes through the forest but bringing power to remote rural communities in a country where access to electricity has risen from just 30 per cent of the population in 2000 to nearly 100 per cent today. Bill Laurance and colleagues (e.g., Laurance and Burgués Arrea, 2017) document the extraordinary rate of road building in areas that have until recently been virtually untouched by modernism, such as the interior Amazon and the Congo Basin. Roads tend to bring uncontrolled and unsustainable development; many people are familiar with satellite images showing "fishbones" of deforestation spreading out from road building in the Amazon; stark though these are they miss the wider impacts from increased hunting and poaching rates that come along with improved access.

In *The New Silk Roads*, Peter Frankopan (2018) has highlighted the extraordinary rate of expansion in road, rail and energy links in Central Asia, another region that has remained relatively isolated for decades, and other areas of China's influence. China's Belt and Road initiative involves over a thousand projects in forty-nine countries; in 2015 the China Development Bank announced it had reserved US$890 billion for this purpose, dwarfing funding for conservation. Arguments against such developments in pristine environments, which predominantly come from conservationists living in countries with a highly effective road system, generally receive short shrift from their colleagues struggling with dirt roads and poor communication. Huge trackless areas are going to become increasingly rare.

All of these things represent real challenges to conservation. As we've said, we need to brace ourselves for further losses. But at the same time, development doesn't automatically mean disaster. Roads in forests do not have to mean deforestation: controversial roads like the one through part of Sangay National Park in Ecuador and Braulio Carillo National Park in Costa Rica have caused far less damage than predicted for instance, although these are admittedly rare examples. Economic development does not automatically mean more land is needed; urbanisation is leading to land abandonment in several areas. In Croatia, as in many of the former communist countries of Europe, the ongoing march of the population to the coast, often linked to employment in the tourism industry, is a factor in the steady annual growth in forest cover, which is evident into rural areas along with abandoned farms and a rapidly aging population.

Government policy is set gradually, by individual governments but also influenced internationally at places like the IUCN World Conservation Congresses and World Park Congresses (Dudley et al., 2014); or the periodic global targets set by the CBD; or by apparently unrelated incidents like decisions about land-based carbon sequestration or rights of indigenous peoples. Are these cumbersome and slow-moving processes enough to address the multiple, fast-moving challenges that we will face in the future? All the signs at the moment are that at least within Europe and North America any more ambitious vision is being lost to petty infighting and increased xenophobia. And as we work on this chapter at least five Latin American countries are in political turmoil, along with several in the Middle East, there are rising tensions in India, riots in Hong Kong and many African nations continue to lurch from crisis to crisis. The major powers seem to be gearing up for a conflict. It doesn't seem like a good period to make grandiose plans to protect major parts of the planet. But then, it probably never has been a good time and there is a sizable part of the global population that does indeed take these issues seriously, and some new and encouraging alliances are emerging.

What we need to do

In the last section of this book we lay out some proposals. It is not exactly a blueprint; there is still so much we don't know about what will happen in the future that it would be premature to be too exact in prescriptions.

Our ideas sit within some broad themes. Particularly on environmental issues, governments seldom lead but are dragged along by pressure from public, industry and non-governmental organisations. Success will depend on a global movement of people. Not just earnest birdwatchers and green advocates, but people in industry, indigenous peoples' groups, schools and religious organisations; a planetary awakening to the reality of climate change and biodiversity loss also gives us an opportunity to develop a far broader social movement. Action needs to take place at all levels. Management of vast ecosystems like the tropical forests and the deep oceans is critical, but climate stabilisation will also entail work on a very local level: from caring about individual butterfly species to caring about the boreal forest.

We group our ideas under six main headings:

1   From protected areas to area-based conservation;
2   From management to stewardship;
3   Recognising all the benefits;
4   Connecting through the mosaic;
5   Focusing on delivery; and
6   Restoring the Earth.

From protected areas to area-based conservation

Large, government-run protected areas have probably already reached, or almost reached, their limit on land in many countries, although there is still a lot to play for in other countries and in the ocean. Utopian international targets that assume every nation can set aside half its land will either be ignored or paid lip service to, or even worse create a backlash. We can also barely find the resources to manage what we have already; setting up a mass of new "paper parks" would be less than useless, creating the impression of protection without the means to achieve anything useful.

This is not to say that protected areas are irrelevant. Protected areas are almost always created for a reason. While a few may have been set up in areas managed sustainably by local people, many more have been established precisely because traditional management processes have broken down, either because the societies are changing or due to outside pressures. The idea that communities will invariably be able to manage their lands and water in harmony with nature is naïve. And we do need new protected areas, including some spectacularly large ones in places that have the space. But the idea that we can or should seek to set aside over half the planet in traditional protected areas is unfeasible.

Fortunately, as we have outlined in previous chapters, we don't have to; there are many more tools available or becoming available, which will likely be more attractive to governments and people. None of them are without their challenges. OECMs are still very new, barely understood even within the conservation community and needing a huge input of work and resources to become properly operational. Other approaches, like Areas of Connectivity Conservation, are even less developed, and the plethora of different approaches beyond these is in many cases almost unknown. Such areas can vary from having management systems that are virtually identical to those in fully protected areas (although they may have different objectives) to types of management that only conserve some elements of biodiversity but play an important role in buffering or supporting protected areas. An effective, but less strict, conservation approach may be more useful than an ineffective protected area.

Working outside strictly protected areas is also essential if healthy populations of endangered species are to be rebuilt. For example, in Nepal, where tiger conservation has been relatively successful, protected areas are already reaching their carrying capacity. Nepali experts calculate that the population is already at 98 per cent of optimum inside the protected areas, but only 60 per cent outside protected areas and 73 per cent in corridors, meaning that future population expansion needs to take place in the wider landscape.

What this also means is that the distinction between what is and what is not "area-based conservation" may become blurred, at least until we learn better how all the different elements fit together into an overall landscape approach. Potential management options that can support all or some biodiversity include various forms of customary management such as sacred natural sites; locally managed marine areas, government no-take zones and codes of practice; various forms of second and third party certification, forest reserves, ecosystem service set-asides, military training grounds, low input or organic agriculture, low-level grazing and hunting reserves; some recreational parks; and temporary measures such as the protection of breeding or over-wintering sites (Dudley and Courrau, 2008). None of these are protected areas. Not all of them are even OECMs (in fact decisions on this will need to be made on a case-by-case basis), but they all fall into a spectrum of being potentially useful for conservation. The extent to which they can or should be included as contributions to the wider vision behind this book will become clearer over the next few years.

Some of these options will be controversial. There is for instance a passionate debate about the extent to which commercial forest use can support biodiversity (e.g. Zimmerman and Kormos, 2012; Putz et al., 2012). However, there is little doubt that carefully managed forests provide valuable buffer and linking habitat and are increasingly being integrated into broader landscape-level conservation strategies.

As yet, most conservation planning approaches do not have the capacity to integrate many different types of area-based conservation, with different levels of significance, governance systems and management approaches, into a coherent whole. But the opportunity is there, and we need to seize it while we can. If we take the full suite of options into account, conserving half the earth suddenly becomes much less of a fantasy (Dudley et al., 2018).

From management to stewardship

Protected areas already cover an unprecedented area of the planet, and they need to expand further. Plus, we can expect to see a whole string of OECMs, connectivity areas and other sites that are, in some cases quite abruptly, taking on new responsibilities for biodiversity conservation. As well as new opportunities, we are creating huge new responsibilities. The present "thin green line" of protected area managers and rangers have a massive task ahead. So too do indigenous peoples devoting their land to sustainable uses, along with sympathetic foresters, fishers, ranchers, tourism operators and company executives. Many people will be learning new skills. Others, with successful, long-established management systems such as many indigenous peoples' groups, will have to adapt these, sometimes radically, as climate changes. Many are already quietly doing so today, while the political classes argue about whether or not climate change is even happening. "Stewardship" refers to "efforts to create, nurture and enable responsibility in landowners and resource users to manage and protect natural and cultural resources" according to the WCPA specialist group on privately protected areas and nature stewardship. As more people necessarily get involved in area-based conservation, we can expect to see an increased movement from management to stewardship.

This is not to imply that we should be moving away from professionalisation – quite the reverse. Changes will involve not only a massive capacity building effort but perhaps more importantly building of a cadre of dedicated individuals around the world, in touch with each other and working together. Too many protected area systems regard a ranger post as a dead-end job, investing little in their staff in terms of either equipment or training, often exposing them to considerable hardship and danger and not surprisingly failing to get optimum returns as a result. The emergence of on-the-ground and distance learning modules aimed at rangers and managers, a widening portfolio of best practice management guides and bodies such as the International Ranger Federation all help address this need. But we still see too many protected areas staffed by disheartened and poorly provisioned staff doing a decent job in trying circumstances.

Efforts need to go further and build similar associations and similar expertise in other communities of practice that will increasingly become part of the delivery mechanisms for area-based conservation: indigenous peoples' organisations, groups of industry people managing areas of their land for conservation, initiatives linking religious organisations that own land, and more. Conservation needs to be seen as a profession of equal importance to healthcare and the law; conservation areas, just as with hospitals and law courts, are nothing without effective staff.

Furthermore, area-based conservation only works in the long run if a critical mass of people supports a particular site. For a long time, conservation relied on governments to set up, administer and protect areas. This is now seen as inadequate; governments are not necessarily to be trusted, disaffected local communities will eventually undermine the site, and many interest groups are simply left out of the debate. A global approach to retaining functioning ecosystem cannot therefore rely on working with our conservation soulmates alone; we will need to be working with stakeholder groups who we do not know, who we may disagree with on some points and who may distrust us. There will be some failures and setbacks, but also important new opportunities.

Indigenous people are particularly relevant stakeholders and rightsholder groups here and are increasingly taking control of the governance of both existing protected areas and sites newly recognised for their conservation values (Artelle et al., 2019). We are wary of non-indigenous conservationists making sweeping generalisations about how indigenous people are going to act in their own territories. But in many cases, conservationists and traditional rightsholders of territories have the same basic objectives in mind for a particular piece of land or water. Rather than disagreeing as has too often been the case, we need to be forming strategic alliances with clear objectives that all parties agree with. Indigenous people, who make up 5 per cent of the global population mainly in Asia, lay claim to at least a quarter of the land surface; some estimates are far higher. Whatever well-meaning UN resolutions might say, holding this territory is going to be a struggle over the next few decades. Several governments in Latin America have stated bluntly that it is simply inequitable for such a small number of people to control so much land, most recently in Brazil. (The same arguments are generally not applied to the rich.) If indigenous people can show that their territories are maintaining wider ecosystem services for the nation or global community, their rights as custodians will be strengthened, but this in turn will imply some constraints on the way that they use the land.

This is easy to say but will not be so easy to achieve. First, we will in many cases have to overcome deep-seated resentments and suspicion, then start the tricky process of agreeing what might be a mutually acceptable way forward. What might work for one generation may not be so acceptable to the next, as indigenous societies are often in a very rapid process of change. The associated negotiations will doubtless be lengthy and complex, and not made any easier by some significant broken promises of the past, but as awareness of climate change and water security issues increases – as it will – these options will become increasingly important.

But we need to be talking with other groups as well, some of whom have been traditional enemies of the conservation movement. Mining companies often have huge areas of land that they own but do not manage. Timber companies can provide linking corridor habitat even if their management destroys much of the original biodiversity. Ranches can contain immensely rich grassland and savannah ecosystems if managed with respect for nature. Religious groups are often amongst the largest landowners in countries and should, in theory, be sympathetic to the conservation cause. All these have links to area-based conservation, and we can point to occasional examples, but this certainly hasn't come to scale as yet.

Recognising all the benefits

As noted, protection of biodiversity in all its forms is only one of the objectives of area-based conservation; maintenance of vital ecosystem services is another, one which some analysts believe needs even more space. We have spent years arguing that many protected areas provide these values, often effectively "free of charge" for governments, business and communities (Stolton and Dudley, 2010). But in the new world of area-based conservation, ecosystem services are also another reason to set aside other areas of land and water, perhaps most significantly for their benefits to climate stabilisation, disaster risk reduction and water security, but with multiple associated benefits that have been examined in Chapter 9.

This will mean interacting with a completely different set of stakeholders. Carbon markets offer a huge potential, particularly as the Paris climate agreements of 2015 are shown to be failing and when ordinary citizens start to demand that the companies they purchase from demonstrate a more responsible approach to climate mitigation strategies. Water security will be important too, especially with the municipalities building huge new cities in Africa, Asia and elsewhere, often with only a fairly vague idea of where the necessary water supplies are going to be sourced. It will mean talking with those responsible for disaster planning and disaster risk reduction strategies and with others responsible for food security, health and the overall wellbeing of citizens. The role of area-based conservation in delivering key Sustainable Development Goals will also be important.

These conversations will not necessarily be easy. Although it is possible to point to many examples of successful use of ecosystem services, there are also huge vested interests pushing hard engineering solutions and saying that natural solutions will not work. Clear advice, examples, links with professional associations, champions and where necessary advocacy will all be needed. User groups, for examples of municipalities drawing on water from protected ecosystems to ensure supply, need to be established to help other towns and cities that are struggling to stabilise their water supply. Just as important as saying when area-based conservation can supply a demand is to be clear when it cannot, and the limitations inherent; one false promise can undo years of positive work.

At a global level, it is also important to work out how the costs can be shared, particularly in the case of ecosystem services that originate in one country but support others. The weather-defining role of vast forests like the Amazon and the Congo are clear examples; many of the agricultural advantages accrue to countries far away and some cost-sharing may be needed to ensure that these ecosystems remain in place.

Despite the huge values that the world already draws from ecosystem services, at present we still barely know where they occur, how much they provide and what needs to be done to maintain them. Building an understanding and respect for ecosystem services is a critical priority for the next few years.

Connecting through the mosaic

As noted, many protected areas exist as islands in a transformed landscape; reconnecting these into a larger functioning ecosystem is an urgent need, albeit one that is often hard to achieve in practice. Connectivity between all types of area-based conservation is important to prevent genetic isolation, maintain the spectrum of habitats needed by mobile species and, if necessary, enable a series of small reserves to "simulate" larger natural areas by allowing species to mingle (Hilty et al., 2019). Connectivity implies that species can move through an area but does not necessarily always require complete continuity of natural habitat, nor does it say much about the speed or regularity of movement. Many species can "jump" some distance over inhospitable territory. There is therefore a difference between "structural" connectivity, which is a direct physical link between suitable habitats and "functional" connectivity where the gap is small enough for species to cross. Species vary dramatically in the distances they will travel outside their native habitat. Some butterflies migrate thousands of miles including across oceans and deserts, while others will barely cross a tarmacked road; the same is true for almost all mobile animals. Plants often disperse much more slowly (Damschen et al., 2006) although species with seeds spread by contact with animals will also be affected by ecological connectivity and plants can disperse fast along free-flowing water courses. It follows that connectivity ideally operates at various scales, for mobile animals like large cats all the way down to slow-growing, slow-dispersing plants.

There are four main forms of connectivity recognized by conservation biologists:

- **Biological corridors** or **ecological corridors**: these are continuous patches of habitat linking two or more native ecosystems. Rivers and hedgerows are two well-known examples found in the general landscape; corridors of native vegetation are sometimes deliberately maintained to link protected areas. Depending on the species, corridors do not have to be totally natural; trees in agricultural areas and grazing pasture can be useful for instance. As corridors become less natural, they will allow passage for fewer species, and their connectivity potential declines.
- **Stepping-stones**: migratory birds, insects and other animals can pass long distances through inhospitable territory but need suitable habitat to rest, hide and feed. Conservation strategies such as the Western Hemisphere Shorebird Reserve Network seek to ensure that there are stopover places for shorebirds travelling twice-yearly along the western coast of North and South America.
- **Buffer zones**: a buffer zone is an area adjacent to a protected area where some management restrictions are in place to support wildlife, such as controls on hunting and intensive development, which will include helping movement of species. Many buffer zones exist in name only, and it is not uncommon to see development right up to the edge of a protected area.
- **Managing the entire landscape for connectivity**: finally, greater connectivity may be achieved by managing an entire habitat with the aim of facilitating inter-connections, for example with a mixture of protected areas and sustainably managed areas.

Natural and semi-natural habitats are particular important of course, but in some cases these can be very small; individual trees or groves can provide enough cover to encourage movement across an otherwise inhospitable landscape for some species, and water edges along rivers and canals can provide passage through cleared areas. Periodic flooding can allow movement of fish and reptiles. Artificial linking habitat, such as underpasses or overpasses crossing roads, artificial nest and roosting sites, fish ladders and perches can all help connect habitats. And although most of the examples given are terrestrial, marine habitats can equally become isolated if the sea bottom becomes degraded and impassable for substrate-living species, or there is lack of vegetation; less is known about connectivity under the surface, but the need is recognised.

Corridors have their critics as well (Bennett, 1999). By increasing accessibility, linkages open new areas to invasive species, novel diseases and new genetic strains of native species, which disrupt local adaptations. Linkages can also facilitate human disturbance such as poaching, fires and pollution. "Hyper-connectivity" (Crooks and Suarez, 2006) refers to an unplanned form of connectivity, whereby species move around the world facilitated by human actions, in ship ballast, on aircraft or attached to the shoes of travellers. In some circumstances it is safer to leave certain populations and habitats isolated, like the offshore islands in New Zealand that maintain populations of ground-living birds that introduced weasels and stoats have driven to extinction on the mainland.

From a management perspective, corridors are also sites of vulnerability because they channel many individuals into a small area. Large-scale shooting of migratory raptors and passerine birds on Mediterranean islands such as Malta undermines conservation efforts for these species throughout the migration route.

Large-scale corridor projects are emerging around the world, with varying degrees of success. Well-known examples include the Australian Alps to Atherton (A2A conservation corridor) initiative; the Alpine Protected Area network in Europe; the Marine Protected Areas Network for the Western Indian Ocean Countries; the Danube Delta and Prut river initiative between Romania, Ukraine and Moldova; and the Yellowstone to Yukon conservation initiative. There are many others.

In the new world of area-based conservation, connectivity is much more than just about linking formal protected areas. It means understanding how a mass of different types of area-based conservation fit together to create a coherent whole: government managed protected areas, ICCAs, private initiatives, OECMs, sustainably managed forests and perhaps a host of other small-scale initiatives. In most cases there is unlikely to be one perfect answer; the overall needs and opportunities will be decided step by step and may well change over time as climate shifts. Most conservation planners have little experience in this kind of multi-layer planning and simple guidebooks do not, as yet, exist. Working with multiple stakeholders also provides additional impetus to the need for far greater participation; if a conservation plan involves fifty different landholders (and it might involve many more), then these people all need to be part of the planning process if it is to succeed. We have many new opportunities, but they come with many new responsibilities as well.

Focusing on delivery

It might be a truism that protected areas only work if they are effective, but this is so often ignored in practice that it is worth stressing. Although the act of declaring a protected area is itself significant, once stresses and pressures emerge any area-based conservation initiative needs resources, knowledgeable and passionate people and clear objectives if it is going to survive.

Improving management effectiveness involves many things. Good governance is often the key. Increasing resources is important, and we have mentioned some of the ways that can be done in previous chapters, although we stress that a reasonable amount of government support (in terms of legislation, policy, avoiding perverse incentives and of course funding) remains essential in most cases. Equally, many privately protected areas and community conserved areas work in isolation, with little support from conservation professionals and few resources to fulfil their objectives. If we are to see a major increase in area-based conservation around the world we need to support this with thousands of well-trained, well-resourced, well-provisioned site-based staff/stewards who clearly understand and contribute to the conservation objectives of the area.

More generally, a focus at site and institutional level in genuine lesson learning and capacity building remains key. Having worked on management effectiveness for two decades we remain committed to the principle of regular assessment and self-reflection but are also aware that far more needs to be done in terms of using assessments in adaptive management; all too many assessments are done as paper exercises to satisfy a government's department or donor agency without being genuinely applied to improve management. Standards and ways of measuring these, such as the IUCN Green List of Protected and Conserved Areas, can help provide a framework against which to measure improvement. The fact that we will now have all kinds of other approaches within the area-based conservation estate makes these changes all the more necessary. New technologies, such as automatic monitoring systems, along with new approaches including crowd-sourced data collection and automated identification systems will make monitoring far more effective in the near future; a mobile phone that can play all the relevant bird calls has revolutionised our own ability to identify birds in new environments and this is just the very beginning of what is becoming possible.

The increase in management effectiveness assessments is a positive step, but we still need to know far more about the links between management and outcomes, about whether species and ecosystems really are being effectively managed through area-based conservation. New technology, greater use of camera traps and AI, and community sourcing of information can all help. Important gaps at the moment include knowledge of conservation outcomes in protected landscapes and seascapes, and under a range of governance types (Dudley and Stolton, 2016).

Delivery also means knowing what we have, and what we could have. The World Database on Protected Areas is a globally important resource, but it is invariably short of funds and struggling to keep pace with changes in the conservation estate. As OECMs and other management strategies become more common these problems will multiply. And as new partners become involved, we need to ensure that they know they can be part of a global system. If someone sets aside a piece of woodland as a nature reserve, how do they tell anyone else, or become part of a wider network? At the moment reporting relies heavily on governments who often cannot be bothered to report anything but their own areas, leading to knowledge gaps, and in turn gaps in opportunities for building more robust systems. We will need streamlined methods of reporting.

Restoring the Earth

The vision outlined in this book is no longer just about leaving space for nature but also creating, or more accurately "recreating", space for nature. In many regions we have already lost so much that simply holding on to what we have is no longer enough. Some restoration efforts, such as clean-up of agricultural pollutants or ocean plastics, need to take place globally and will be unlikely to focus explicitly on area-based conservation. But bringing restoration into the conservation arena has sometimes been a struggle; people have argued that restoration is a sign of failure, that natural ecosystems can never be restored – once gone, lost forever – and that the sole focus of efforts should be on conservation.

In fact, many of the current "natural ecosystems" have been cleared at some time in the past, particularly in the Americas where the introduction of European diseases caused a colossal population collapse and subsequent regrowth of what the early European settlers believed was a natural landscape (e.g. Flannery, 2001). Today land degradation is also leading to increasing agricultural abandonment and the subsequent regrowth of natural vegetation. Initiatives like the Bonn Challenge are attracting governments to commit to large-scale reforestation.

Ecologists will be quick to point out that these regenerated forests or rangelands are seldom exactly the same as what came before and are often less diverse (or at least missing a proportion of the original species). This is the reason why we fight so hard to keep the most pristine natural environments from being destroyed. But getting back a diverse and functioning ecosystem is usually possible, even if it is not exactly the same as the one there before degradation took place. The potential for meaningful land restoration is enormous, and the planned global Decade of Restoration, from 2021 onwards, will hopefully provide an important opportunity to focus attention on rebuilding ecosystems that have been lost or degraded. Already abandoned by other uses a proportion of these will be suitable contributors to area-based conservation targets. Experiments in large scale reforestation in South Korea (Korea Forest Service, undated) and in Guanacaste in Cota Rica (Janzen, 2000) show what is possible; in the latter case the ecosystem is so rich that it has been declared a national park.

Clawing back land from the desert in Saudi Arabia and Kuwait

It is extraordinary, and encouraging, just how much will come back with a little effort, even in places that might appear to be wrecked forever. Large parts of the Arabian Peninsula have been overgrazed for so long that people have no idea the environment could amount to anything else. Indeed, some areas are too dry to support much except sand and rocks. But restoration efforts in association with staff at Kings Botanic Garden in Perth, Australia, have created dramatic new vegetation patches outside Riyadh, places where suddenly there are birds singing and insects buzzing from flower to flower (Arriyadh Development Authority, 2013). Admittedly, this was achieved using the sophisticated restoration techniques developed by a research establishment which is itself based in an arid part of the world and has decades of experience. But restoration can often also take place just by taking away the pressures that have caused degradation. Kuwait suffered some of the worst environmental pollution ever seen when Saddam Hussein set fire to the oil wells at the end of the first Gulf War; when I visited, much of the country was literally covered with a centimetre-thick layer of tar, today often hidden by a thin coating of sand (a massive restoration programme is now underway). The demilitarised zone between Kuwait and Iraq is a classic no-go area; we had to obtain special permission and negotiate over coffee and dates with the army posts found at regular intervals along the boundary road. But once inside this fenced and desolate area, the desert vegetation was dramatically different from that found outside; in the absence of most grazing animals the twenty years following the war had seen a reappearance of plant species rare in the rest of the country and a consequent reduction in erosion and appearance and more permanent vegetation patches. Given even half a chance nature will try to reassert itself. **ND**

To some extent, this will involve restoration within protected areas and OECMs (Keenleyside et al., 2012). But this is not the end of the story. Researchers have recently pointed out that one of the quickest ways of addressing climate change is through a really major global reforestation effort (Bastin et al., 2019), and govern-ments are starting to make pledges. There is huge potential. But such a strategy also contains inherent dangers; many people are still living on "degraded land", and their views need to be taken into consideration. We got involved in assessing a plantation project in Borneo many years ago where the government had handed a company a large area of "degraded" land for establishment of timber plantations without mentioning the ninety-plus villages found in the area. Furthermore, grass-land and savannah habitats are arguably being lost even more quickly than forests, and poorly planned reforestation efforts could undermine these ecosystems. They could also undermine forest ecosystems if degraded forests are replaced by fast-growing plantations of exotic trees. So, while we believe that restoration is a critical part of the overall picture, it needs to be approached with considerable caution. Restoration may be as much about improving the carbon storage in grasslands as it is about planting many more trees, depending on where it is taking place. Perhaps even more important in the long term is ocean restoration and the potential of blue carbon: the restoration of the ocean to provide additional carbon sequestration (Howard et al., 2017).

Finally, some blue-sky thinking

Other changes will influence the potential for area-based conservation. There are around 570 million farms worldwide, of which 410 million are less than a hectare in size and 475 million less than 2 ha, supporting over a billion farmers but occupy-ing less than 12 per cent of agricultural land (Lowder et al., 2016). Current trends suggest that many, perhaps most, of these small farms will disappear (Dudley and Alexander, 2017). Many will be consolidated into larger, more profitable enterprises, but others, on unproductive land or in harsh conditions, could well simply be abandoned, like the once extensive upland farms of Vermont that are now covered in forest. While projections currently suggest that additional areas will need to be brought into agriculture due to rising populations, coupled with increasing consumer demands (Taylor, 2011), this will be heavily influenced by dietary choices. Such changes would have profound cultural impacts, many of which we deplore, but they are also likely to free up additional land for ecosystem services or other lower impact uses. Similar issues arise with forests; while these continue to be lost rapidly at a global scale, in several countries of Europe and in parts of North America they are rapidly regenerating.

In particular, the future of livestock production (and the additional land needed to produce livestock feed) is critical. Some meat substitutes already taste virtually like meat; in a few years they will be indistinguishable and less costly in many ways. Insect protein will become increasingly common and far more efficient to produce (Dobermann et al., 2017). The number of vegetarians and vegans is growing rapidly; a new generation of plant-based products that do not sacrifice taste or nutrition could transform large parts of the food system in just a few decades; another generation might regard it as generally unnecessary to keep and kill animals for food when there are indistinguishable and cheaper alternatives. Such changes would again have enormous social and cultural impacts on those areas of the world that rely largely on livestock production, but they undoubtedly have a significant potential to reduce the demand for land. Similarly, we don't know what impacts if any the new types of genetically modified (GM) crops will have on land availability, nor whether more traditional ways of crop-breeding will impact land use requirements (Gilbert, 2016). A widespread move towards organic agriculture would increase the amount of land needed for production, which is the reason some conservationists oppose organic, but farms like these can provide a positive benefit to biodiversity conservation and other ecosystem services and are already incorporated into some protected landscapes (Stolton, 2005).

If the worst predictions of climate change specialists are right, the world will be profoundly different for a long time, and incidentally will probably be facing societal breakdown on a level not seen before. There is still little sign that some of the major players will change their ways in time, still huge amounts of funding pouring into disinformation campaigns and politicians growing rich on playing the climate denial card. The current so-called migrant crisis will escalate dramatically if places become unliveable. It may be that we are indeed leaving future generations with a planetary crisis to sort out. What should they do then? Are we just wasting our time?

We have barely started to think about how to respond to worst case scenarios as yet. If ocean acidification really is as bad as scientists fear, do we simply abandon the world's coral reefs or do we keep some alive through artificial liming, in much the same way as Sweden and Norway countered freshwater acidification thirty years ago? Similar questions exist for tropical forests faced with drying out, or for species that are threatened by novel diseases, loss of habitat or ecological disruption. Under these conditions, keeping ecosystems alive will literally be a matter of life and death. But given that we are presenting the future with an unprecedented set of problems, it seems only fair that those who care try to start looking for some innovative responses. Giving up will by then no longer be an option.

The future of Snowdonia National Park

To return to our home patch. Snowdonia National Park stretches over 2,170 km² from close to the coast of North Wales, south to the River Dyfi. Snowdonia is a protected landscape, IUCN Category V, an area of mountains, lakes, woodland, conifer plantations and small settlements with a coastline bordering the Irish Sea. A series of ancient volcanoes form the basis of a mountain chain, the so-called ring of fire, and their eroded dome contains some of the oldest surface rocks on the planet. Heavy rainfall and high humidity define the native Welsh oak woodlands as temperate rainforest, with globally important moss and liverwort populations – the Atlantic bryophytes. With a fiercely independent population, the Welsh language still predominates. Historically the area has been dedicated to upland farming, which has over time simplified from mixed farming to an overwhelming focus on sheep. Three generations have been kept afloat on subsidies from the European Union and now face an uncertain future. Abandoned quarries litter the hillsides, predominantly slate but also metal mines, mainly magnesium, silver and gold. After the First World War and until the 1970s, the government's Forestry Commission had sweeping powers to purchase land to provide a strategic supply of timber. Huge areas are covered with exotic spruce and pine plantations, many containing the ruined farmsteads of people who were forced off the land to make way for plantations. The main income today comes from tourism, although farming and forestry still dominate land-use decisions. Snowdonia, like other UK national parks, just about scraped into meeting the IUCN definition of a protected area when we re-examined all the UK protected area categories in 2014 (Crofts et al., 2014). It is the part of the planet we know best, and we've walked for years throughout the area, returning repeatedly to favourite mountains and rivers, camping on the tops, canoeing down the estuaries and trying to understand the complex nature of the park.

What is the future of a protected area like Snowdonia? Recently the Welsh government made noises about radically changing its status, although quickly backed down in the face of opposition. Nature conservation is not particularly successful; upland birds are declining for reasons not fully understood. There have been no recent extinctions, but numbers continue to fall, and populations of species like the hare are a fraction of what they were in the nineteenth century, having been decimated by hunting and land-use change.

Some things have returned; the red kite has bounced back from a rump population and now regularly swoops over our house, and the pine martin has recently been re-introduced. Farming will continue to decline and rewilding projects are increasing in significance, although still fiercely controversial. Snowdonia could well become more "natural" in the next few decades, as upland farming gradually shrinks, and remaining government support is aimed more towards ecosystem services. Land ownership will likely change, passing to people who have a different vision of what Snowdonia represents. But none of this is certain. The hills could instead be planted with rapidly growing energy crops, or land ownership consolidated to a few rich people and continue much as it is at the moment; there could be depopulation or repopulation. **ND/SS**

In this book we have laid out an optimistic vision of what could and should happen: a huge expansion of those areas of the world managed consciously to maintain nature and ecosystems service, coupled with the maintenance and protection of vulnerable human societies. We've focused a lot on the nuances of language: "area-based conservation" rather than only "protected areas", "stewardship" rather than just "management". If we want what will amount to a global revolution in the way in which we approach land and sea, we need to shift the axis of the debate in subtle but fundamental ways. But the future remains uncertain, as always, and some of the signs are that things are going to get much worse before they get better, if indeed they ever do get better. It is a time for cautious optimism but also stamina and courage.

References

Arriyadh Development Authority. (2013) *Preservation, Rehabilitation and Development of Native Plant Cover in Arriyadh Province. Phase 1: Background to Nature Conservation Strategy for Arriyadh Province.* Joint report, Kings Park and Botanic Garden Perth, Royal Botanic Garden Edinburgh and Equilibrium Research

Artelle, K.A., Zurba, M., Bhattacharrya, J., Chan, D.E., Brown, K., et al. (2019) 'Supporting resurgent indigenous-led governance: A nascent mechanism for just and effective conservation'. *Biological Conservation*, vol. 240, no. 108284. Doi:10.1016/j.biocon.2019.108284

Bastin, J.F., Finegold, Y., Garcia, C., Mollicone, D., Rezende, M., et al. (2019) 'The global tree restoration potential'. *Science*, vol. 365, pp 76–79

Bennett, A.F. (1999) *Linkages in the Landscape: The Role of Corridors and Connectivity in Wildlife Conservation.* IUCN, Gland, Switzerland

Crofts, R., Dudley, N., Mahon, C., Partington, R., Phillips, A., et al. (2014) *Putting Nature on the Map: A Report and Recommendations on the Use of the IUCN System of Protected Area Categorisation in the UK.* IUCN National Committee, London

Crooks, J.A. and Suarez, A.V. (2006) 'Hyperconnectivity, invasive species, and the breakdown of barriers to dispersal'. In: Crooks, J.A. and Sanjayan, M. (eds.) *Connectivity Conservation*, Cambridge University Press, Cambridge, pp 451–478

Damschen, E.I., Haddad, N.M., Orrock, J.L., Tewksbury, J.J. and Levey, D.J. (2006) 'Corridors increase plant species richness at large scales'. *Science*, vol. 313, pp 1284–1286

Dobermann, D., Swift, J.A. and Field, L.M. (2017) 'Opportunities and hurdles of edible insects for food and feed'. *Nutrition Bulletin*, vol. 2, no. 4, pp 293–308

Dudley, N. and Alexander, S. (2017) 'Will small farmers survive the 21st century – and should they?' *Biodiversity*. Doi:10.1080/14888386.2017.1351397

Dudley, N. and Courrau, J. (2008) 'Filling the gaps in protected area networks: A quick guide for protected area practitioners'. In: Ervin, J. (ed.) *The Nature Conservancy*, Quick Guide Series, Arlington VA

Dudley, N. and Gordon Clarke, J. (1983) *Thin Ice*. Marine Action Centre, Cambridge

Dudley, N., Groves, C., Redford, K.R. and Stolton, S. (2014) 'Where now for protected areas? Setting the stage for the 2014 world parks congress'. *Oryx*. Doi:10.1017/S0030605314000519

Dudley, N., Jonas, H., Nelson, F., Parrish, J., Pyhälä, A., et al. (2018) 'The essential role of other effective area-based conservation measures in achieving big bold conservation targets'. *Global Ecology and Conservation*, vol. 15, p e0024

Dudley, N. and Stolton, S. (2016) 'Protected area diversity and potential for improvement'. In: Joppa, L.N., Baillie, J.E.M. and Robinson, J.G. (eds.) *Protected Areas: Are They Safeguarding Biodiversity?* Wiley Blackwell and Zoological Society of London, Chichester and London

Flannery, T. (2001) *The Eternal Frontier: An Ecological History of North America and Its Peoples.* William Heinemann, London

Frankopan, P. (2018) *The New Silk Roads: The Present and Future of the World*. Bloomsbury, London

Gilbert, N. (2016) 'Cross-bred crops get fit faster'. *Nature*, vol 513, p 292

Hall, D.K., Giffen, B.A. and Chien, J.Y.L. (2005) 'Changes in the Harding Icefield and the Grewingk-Yalik Glacier complex'. *62nd Eastern Snow Conference*, Waterloo, ON, Canada

Hilty, J., Keeley, A.T.H., Lidicker Jnr, W.Z. and Merenlender, A.M. (2019) *Corridor Ecology*, 2nd edition. Island Press, Covelo, CA

Howard, J., Sutton-Grier, A., Herr, D., Kleypas, J., Landis, E., et al. (2017) 'Clarifying the role of coastal and marine systems in climate mitigation'. *Frontiers in Ecology and the Environment*, vol. 15, no. 1, pp 42–50. Doi:10.1002/fee.1451

Janzen, D.H. (2000) 'Costa Rica's Area de Conservación Guanacaste: A long march to survival through non-damaging biodevelopment'. *Biodiversity*, vol. 1, no. 2, pp 7–20

Keenleyside, K., Dudley, N., Cairns, S., Hall, C. and Stolton, S. (eds.) (2012) *Ecological Restoration for Protected Areas: Principles, Guidelines and Best Practice*, Best Practice Protected Area Guidelines no. 18. IUCN, Gland, Switzerland

Korea Forest Service. (undated) *Lessons Learned from the Republic of Korea's National Reforestation Program*. Federal Ministry for the Environment, Nature Conservation, Building and Nuclear Safety and Convention on Biological Diversity, Seoul and Montreal

Laurance, W.F. and Burgués Arrea, I. (2017) 'Roads to riches or ruin?' *Science*, vol. 358, no. 6362, pp 442–444

Lowder, S.K., Skoet, J. and Raney, T. (2016) 'The number, size, and distribution of farms, smallholder farms, and family farms worldwide'. *World Development*, vol. 87, pp 16–29

Lynn, K., Daigle, J., Hoffman, J., Lake, F., Michlle, N., et al. (2013) 'The impacts of climate change on tribal traditional foods'. *Climatic Change*, vol. 120, pp 545–556

National Park Service. (2019) *2019 Seabird Die-Off*. www.nps.gov/subjects/aknatureand science/commonmurrewreck.htm. Accessed 16 November 2019

Putz, F.E., Zuidema, P.A., Synnott, T., Peña-Claros, M., Pinard, M.A., et al. (2012) 'Sustaining conservation values in selectively logged tropical forests: The attained and the attainable'. *Conservation Letters*, vol. 5, no. 4, pp 296–303

Stolton, S. (ed.) (2005) *Organic Agriculture for Biodiversity: Current Contributions and Future Possibilities*. Proceedings of the Third International IFOAM Conference on Biodiversity and Organic Agriculture: Nairobi 2004. International Federation of Organic Agriculture Movements, Tholey Theley, Germany

Stolton, S. and Dudley, N. (eds.) (2010) *Arguments for Protected Areas: Multiple Benefits for Conservation and Use*. Earthscan, London

Taylor, R. (ed.) (2011) *WWF Living Forests Report. Chapter 1: Forests for a Living Planet*. WWF, Gland, Switzerland

Zimmerman, B. and Kormos, C. (2012) 'Prospects for sustainable logging in tropical forests'. *Bioscience*, vol. 62, no. 5, pp 479–487

INDEX

Note: page numbers in **bold** indicate a table on the corresponding page.